JIMD Reports
Volume 5

SSIEM

JIMD Reports –
Case and Research Reports, 2012/2

 Springer

Editor
Society for the Study of Inborn
Errors of Metabolism
c/o ACB
Tooley St 130-132
SE1 2TU London
United Kingdom

ISSN 2192-8304   e-ISSN 2192-8312
ISBN 978-3-642-28095-5   e-ISBN 978-3-642-28096-2
DOI 10.1007/978-3-642-28096-2
Springer Heidelberg Dordrecht London New York

Printed on acid-free paper

Springer is part of Springer Science+Business Media (www.springer.com)

Contents

JIMD Reports
DOI 10.1007/8904_2011_109

Homocysteine Measurement in Dried Blood Spot for Neonatal Detection of Homocystinurias

Ahmad N. Alodaib · Kevin Carpenter ·
Veronica Wiley · Tiffany Wotton · John Christodoulou ·
Bridget Wilcken

Received: 6 June 2011 /Revised: 3 October 2011 /Accepted: 25 October 2011 /Published online: 13 December 2011
© SSIEM and Springer-Verlag Berlin Heidelberg 2011

Abstract Expanded newborn screening (NBS) leads to an increased number of false positive results, causing parental anxiety, greater follow-up costs, and the need for further metabolic investigations. We developed and validated a second-tier approach for NBS of homocystinurias by measuring the total homocysteine (tHcy) on the initial dried blood spot (DBS) samples to reduce the need for further investigation, and investigated newborn DBS homocysteine values in patients with homocystinuria. Total DBS homocysteine was measured in normal newborns, and retrospectively in newborns with established disorders, using liquid chromatography tandem mass spectrometry (LC-MS/MS) with stable isotope-labelled internal standards for homocysteine. Analytes were separated using reverse phase chromatography with a total run time of 3 min. The method was linear over the range of 10–100 μmol/L of tHcy and showed excellent precision; intra-batch CV was 4% and inter-batch precision 6.5%. Comparison of 59 plasma values with DBS for tHcy taken at the same time showed excellent correlation, ($r^2 > 0.97$). The reference range for current neonatal samples was 5.4–10.7 μmol/L ($n = 99$), and for the stored neonatal samples (stored dry, sealed in plastic at room temperature for 10 years) was 1.7–5.5 μmol/L, ($n = 50$), both being normally distributed. The clinical utility of this method was checked by retrospective analysis of stored NBS samples from patients with different forms of homocystinuria, including four different remethylating disorders. All had clear elevations of tHcy.

Communicated by: Georg Hoffmann.

Competing interests: None declared.

A.N. Alodaib · K. Carpenter · V. Wiley · T. Wotton ·
J. Christodoulou · B. Wilcken
Discipline of Paediatrics and Child Health, University of Sydney,
The Children's Hospital at Westmead Clinical School, Locked
Bag 4001, Westmead, NSW 2145, Australia

A.N. Alodaib
Department of Genetics (MBC-03), King Faisal Specialist
Hospital and Research Centre, P.O. Box 3354, Riyadh 11211,
Saudi Arabia

K. Carpenter · V. Wiley · J. Christodoulou · B. Wilcken
Discipline of Genetic Medicine, University of Sydney, The
Children's Hospital at Westmead Clinical School, Locked Bag
4001, Westmead, NSW 2145, Australia

K. Carpenter (✉) · B. Wilcken
NSW Biochemical Genetics Service, The Children's Hospital at
Westmead, Locked Bag 4001, Westmead, NSW 2145, Australia
e-mail: kevin.carpenter@health.nsw.gov.au

V. Wiley · T. Wotton · B. Wilcken
NSW Newborn Screening Programme, The Children's Hospital at
Westmead, Locked Bag 4001, Westmead, NSW 2145, Australia

Abbreviations

AC	Acylcarnitine
C3	Propionylcarnitine
cbl	Cobalamin
CBS	Cystathionine β-synthase
DBS	Dried blood spot
ESI-MS/MS	Electrospray tandem mass spectrometry
LC-MS/MS	Liquid chromatography tandem mass spectrometry
Met	Methionine
MS/MS	Tandem mass spectrometry
MTHFR	Methylene tetrahydrofolate reductase
NBS	Newborn screening
tHcy	Total homocysteine

Introduction

The homocystinurias are a diverse group of disorders comprising one defect in trans-sulphuration, cystathionine β-synthase (CBS) deficiency, and defects in remethylation, principally disorders of cobalamin (cbl) metabolism, cbl C, D, E, F, G defects, defects of folate metabolism, and methylene tetrahydrofolate reductase (MTHFR) deficiency (Fowler 1997).

The clinical presentations are diverse. Homocystinuria due to CBS deficiency is associated with thrombo-embolic events, ectopia lentis, mental retardation, psychiatric disorders, and skeletal abnormalities (Refsum et al. 2004; Yap and Naughten 1998). The remethylation disorders share a somewhat similar clinical presentation with failure to thrive, acute or chronic neurological deterioration, developmental delay, and sometimes seizures, hypotonia, microcephaly, feeding difficulties, stomatitis, and microcephaly. The cbl defects also present with megaloblastic anaemia, although measured serum vitamin B12 may be normal (Digest 2007; Schiff et al. 2011). In the cbl and folate deficiencies, megaloblastic anaemia is a classical clinical feature, and there are different neuropsychiatric abnormalities which can appear in cbl deficiency due to demyelination of peripheral nerves, the spinal cord, cranial nerves, and the brain (Whitehead 2006). Treatment is available for all these disorders, and there seems likely to be a clinical advantage in starting treatment in the newborn period.

Newborn screening (NBS) by tandem mass spectrometry routinely detects pyridoxine non-responsive CBS deficiency and cblC defect, and may be able to detect the other disorders. NBS for the homocystinurias involves initial measurement of methionine (Met) to detect CBS deficiency, in which Met levels are elevated, and propionyl and acetyl carnitines (C3, C2) to detect the cblC defect, and possibly other remethylation defects. Using this first-tier approach sensitivity is poor for CBS deficiency as a whole, and specificity somewhat poor for both (Wilcken et al. 2003).

Measurement of total homocysteine (tHcy) in DBS samples has been used as a second-tier test in screening for CBS deficiency (Matern et al. 2007), and recently as an initial test (Gan-Schreier et al. 2009). Measurement of methionine and methionine:phenylalanine ratios to detect low levels has been thought useful for the detection of remethylation defects (Tortorelli et al. 2010; Turgeon et al. 2010).

In this chapter, we report our findings from measuring tHcy in DBS in normal newborns and, retrospectively, in newborns with established disorders, using an LC-MS/MS method.

Materials and Methods

Reagents

DL-Homocysteine (95% pure) was purchased from Aldrich Chemical Company (Sydney, NSW, Australia). DL-Dithiothreitol was purchased from Sigma Chemical Company (Sydney, NSW, Australia). HPLC-grade methanol was purchased from BDH chemicals (Minto, NSW, Australia). Formic acid (FA) was obtained from Ajax Finechem (Taren Point, NSW, Australia). The isotopically labelled internal standard (Homocystine $-d_8$) for Hcy was purchased from Cambridge Isotope Laboratories (Andover, MA, USA). The filter paper used for the sample collection was grade 903 (Whatman, Kent, UK). Ultrapure water was generated using a Millipore-MilliQ system (Millipore, Kilsyth, VIC, Australia).

Samples

The reference range of tHcy in normal newborn DBS was established using 99 current neonatal samples. A separate range for samples stored dry, sealed in plastic at room temperature was obtained using 50 samples stored for 10 years and analysed by LC-MS/MS. As part of the method validation, 59 DBS samples were obtained from blood samples of homocystinuric patients submitted for analysis of plasma homocysteine. The tHcy determined in these DBS on LC-MS/MS was then correlated with simultaneous plasma sample results. Plasma tHcy was analysed by LC-MS/MS method developed in house, based on the method of Magera et al. (1999). In addition, we measured DBS tHcy in six patients with confirmed CBS deficiency, ten cblC deficient patients, two cblG patients, and one MTHFR patient. These samples were stored at room temperature under dry conditions.

Experimental Conditions

The HPLC analysis of tHcy in DBS was based on the method of McCann et al. (2003) with some modifications. A Waters 1,525 μ Binary HPLC Pump system (Waters Corporation, Rydalmere, NSW, Australia) was used, and separation from the bulk of the sample matrix performed on an Altima C_{18} (150 × 2.1 mm, 3 μm) column (Alltech Associates Australia Pty Ltd, Baulkham Hills, NSW, Australia) equipped with a cartridge guard column. The chromatographic separation was performed using isocratic elution with a mobile phase of methanol:water (30:70, v/v) containing 0.1% formic acid, injection volume 10 μL, flow rate of 200 μL/min, and total run time of 3 min. A Quattro Micro tandem mass spectrometer (Waters Corporation,

Rydalmere, NSW, Australia) was used for the detection and was operated in a positive ion multiple reaction monitoring (MRM) mode. The ion source was operated at 3.5 kV and the temperature of source gas was 110°C. The optimised MRM transitions used for the measurement of tHcy were m/z 136 → 90 for the native species and m/z 140 → 94 for the d4 homocysteine internal standard. The data were analysed using QuanLynx software.

Standards

A stock solution of DL-homocysteine (7.4 mmol/L) was prepared by dissolving 100 mg of DL-homocysteine in 100 mL of 0.02 mol/L HCl. For preparation of standards, the working dried blood spot (DBS) standards were prepared from a healthy control blood sample with added Hcy. The added amounts were 0, 10, 50, and 100 μmol/L of Hcy. Standards were spotted onto NBS cards and left to dry overnight at room temperature. After drying, DBS standards were placed in plastic bags separately and stored at −20°C. Endogenous homocysteine in the sample was estimated by back calculation using least squares regression following analysis of enriched standards.

Sample Preparation

A single 3 mm disc was punched out of the DBS samples and standards in a 1.5 mL microcentrifuge tube Eppendorf (Hamburg, Germany), followed by addition of 20 μL of 10 μmol/L internal standard. Tubes were vortex-mixed gently for 1 min and then 20 μL of 500 mmol/L Dithiothreitol and 100 μL of M-Q water containing 0.1% (w/v) formic acid were added to the tubes. Each tube was mixed for 2 min and then centrifuged at $2,150 \times g$, 40°C for 5 min to remove any particulate matter. Eighty microlitres of supernatant was transferred to the 96-well polypropylene microtitre plate for automated injection.

Results

Linearity was tested using the prepared standard curve over the range of 10–100 μmol/L of tHcy. The minimum correlation coefficient for this standard curve was 0.997 over 21 batches with negligible intercepts observed. The inter- and intra-assay precision expressed as the coefficient of variation was evaluated using a quality control sample of blood spiked with 10 μmol/L Hcy and was 4.0% and 6.5%, respectively ($n = 96$ and 16).

Limits of detection were not formally tested; however, signal to noise was >100:1 at the lowest levels of homocysteine observed (1.7 μmol/L). The results for DBS samples analysed by this method were compared with the

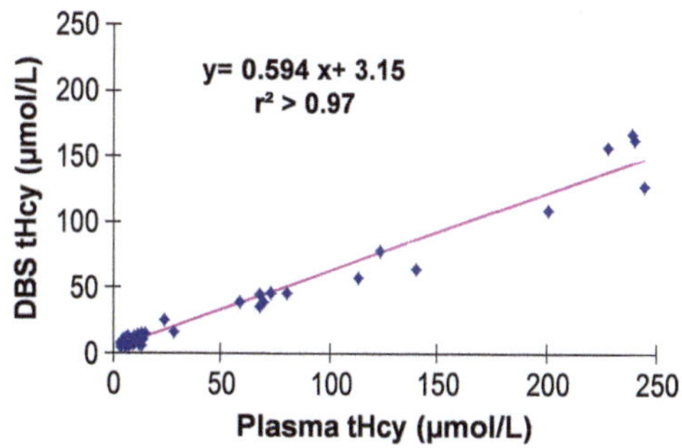

Fig. 1 Regression analysis of DBS results for homocysteine analysed by HPLC-MS/MS and plasma samples taken simultaneously analysed by HPLC-MS/MS. Squared correlation coefficient (r^2)

plasma samples collected at the same time. The linear regression analysis of the results ($n = 59$) showed excellent correlation ($r^2 = 0.97$) (Fig. 1). It is important to note that no attempt was made to create a directly comparable result with plasma since this assay is purely for second-tier NBS testing, and not used for monitoring.

Reference ranges were established for both the current neonatal DBS samples and the stored neonatal DBS samples to evaluate the effect of the storage time. Reference ranges were calculated as (mean ± 1.96 SD) in both stored and current samples. The reference range obtained for current neonatal samples was 5.4–10.7 μmol/L ($n = 99$), and for samples stored for 10 years was 1.7–5.5 μmol/L ($n = 50$). Figure 2 shows the distribution of tHcy concentrations from both the current and the stored normal DBS.

We retrospectively analysed 20 NBS samples from patients with homocystinuria of different aetiologies. They included six patients with confirmed CBS deficiency (one patient had two NBS samples taken), ten with cblC, two with cblG, and one with MTHFR deficiency. This method clearly identified the affected patients in DBS, as all patients had very clear elevations of tHcy. Combined results of those patients are shown in Table 1, and the distribution of tHcy in those patients is shown in Fig. 3.

Methionine:phenylalanine ratios were calculated retrospectively in NBS samples from patients with remethylation defects and ranged from 0.10 to 0.38 with all but one being below the cutoff of 0.22 suggested by Tortorelli et al. (2010).

Discussion

Several LC-MS/MS methods have been reported to determine tHcy in DBS, plasma, and urine (Gempel et al. 2000; Magera et al. 1999). In this report, a stable isotope dilution LC-MS/MS method based on the method of McCann et al. (2003) was developed and validated for the

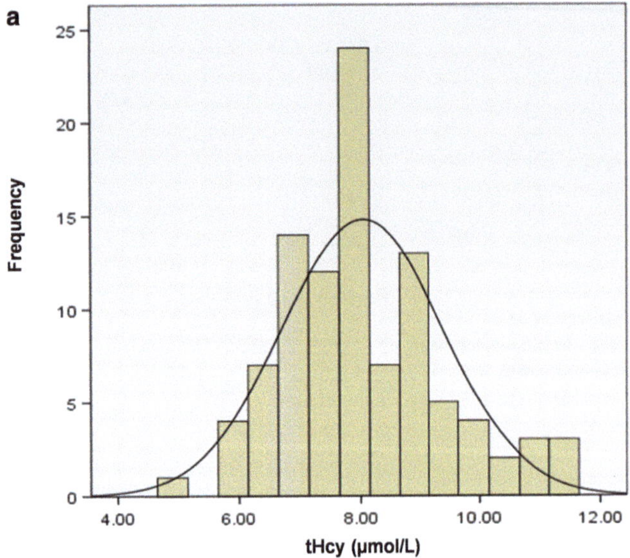

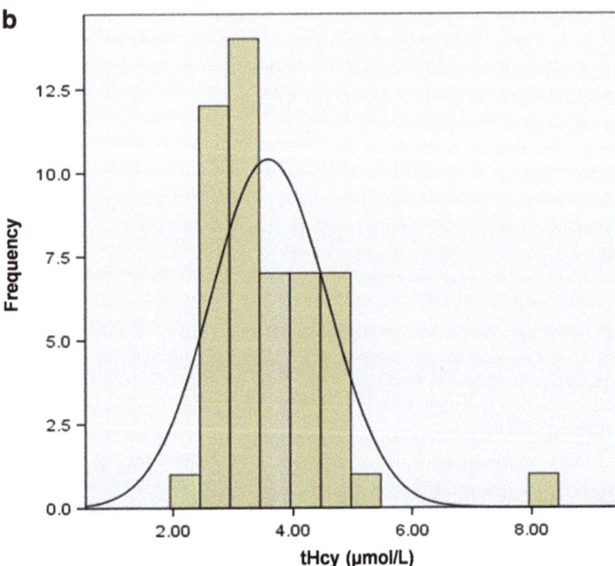

Fig. 2 The distribution of tHcy concentrations from (**a**) current neonatal DBS and (**b**) stored neonatal DBS

quantification of tHcy in DBS samples as a second-tier NBS test for the diagnosis of suspected homocystinuria patients. This method is also suitable for adult population screening for mild hyperhomocystinaemia, taking advantage of its fast separation and detection within 3 min, easy sample collection and delivery to the testing laboratory, and low cost per assay.

This method was validated by evaluating its linearity, precision, and between methods comparison. This method was linear over the range of 10–100 μmol/L of tHcy. The method also showed excellent precision; intra-batch CV was 4%, while inter-batch precision was 6.5%. Our results of comparison of 59 plasma values with DBS for tHcy taken at the same time showed an excellent correlation ($r^2 > 0.97$), but with the previously noted lower values in DBS (Fig. 1).

However, no attempt was made to create a directly comparable result with plasma in this assay since this assay is purely for second-tier NBS testing, and not used for monitoring.

Our results showed that the tHcy concentrations in DBS are clearly lower than in plasma, which is in accord with the previous report of McCann et al. (2003). They suggested that it is likely the concentration of homocysteine in plasma is higher than red blood cells, thus explaining the lower results obtained from whole blood samples.

Since retrospective analysis of stored samples from positive cases was undertaken, reference ranges were established for both the current neonatal DBS samples and the stored neonatal DBS samples to examine the effect of extended storage time. Reference ranges calculated as mean ± 1.96 SD in both stored and current samples were normally distributed (Fig. 2). The reference range determined in the current neonatal samples was 5.4–10.7 μmol/L ($n = 99$), which is in close agreement with published reports (Gan-Schreier et al. 2009). The reference range in stored neonatal samples was markedly lower (1.7–5.5 μmol/L; $n = 50$), illustrating the loss of Hcy on long-term storage, which agrees with a previous report by Bowron et al. (2005). They found that the Hcy concentration in DBS is stable at room temperature for 24 h, with a reduction after 28 days of storage.

In order to evaluate the clinical utility of this method, 20 stored NBS samples from patients with various inborn errors of metabolism causing homocystinuria were retrospectively analysed. This method clearly identified those affected patients with homocystinuria with obvious elevations of tHcy, all of which, despite storage, were above the reference range for current samples (Fig. 3 and Table 1). In terms of the primary markers to trigger second-tier testing, it was clear that elevated Met appeared in the classical homocystinuria patients, while either low or normal Met levels were found in the patients with disorders of homocysteine remethylation. Thus, in our experience, although low Met might be a helpful marker for remethylating defects, it has poor sensitivity. This may be improved by the use of ratios to phenylalanine as suggested by others (Tortorelli et al. 2010; Turgeon et al. 2010), and retrospective review of our data where initial NBS was performed by tandem mass spectrometry supports this with five of the six samples showing a methionine:phenylalanine ratio <0.22, the cutoff used by Tortorelli et al.

On the other hand, propionylcarnitine was usually elevated either primarily or as a ratio with acetylcarnitine in all patients with cblC defect (one of the homocysteine remethylation disorders), while this marker was normal in patients with classical homocystinuria and the other two defects of remethylation, MTHFR and cblG. These observed results are consistent with the literature (Tortorelli et al. 2010; Turgeon et al. 2010).

Table 1 Metabolic data of 19 patients diagnosed with homocystinurias

Disorder (number of cases)	NBS results (assayed in newborn period)		tHcy in DBS (assayed after storage)	Sample storage
	Met(cutoff <5 or >75) µmol/L	C3(cutoff >8.5) µmol/L	Reference range 5.4–10.7 (current) 1.7–5.5 (stored) µmol/L	
CBS (6 cases, 7 samples)[a]	120–470	0.6–3.2	21.2–46.6	4 m[c] to 10 y[d]
cblC (10)[b]	6.3–19.3	6.6–16	11.1–186.3	1 m to 15 y
cblG (2)	16 and 21	2.1 and 5.6	13.5 and 25.8	4 and 16 y
MTHFR (1)	16	2.1	93.4	2 y

[a] The actual Met levels of CBS patients are 120, 150, 190, 190, 250, 350, and 470 µmol/L

[b] The actual Met levels of cblC patients are 6.3, 7.5, 8.1, 10.9, 16.6, 19.3 µmol/L. Four samples were quantified prior to tandem mass spectrometry (MS/MS) use in NBS

[c] *m* month

[d] *y* years

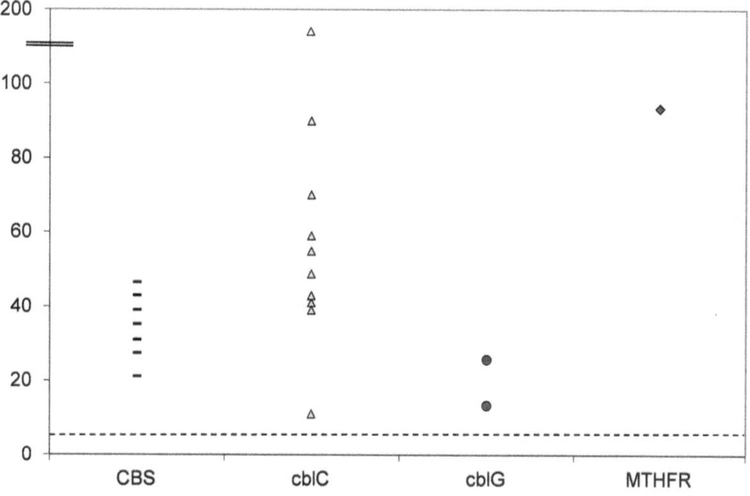

Fig. 3 Distribution of tHcy (µmol/L) in DBS from 19 individuals with a metabolic disorder associated with homocystinuria. *Dotted line* denotes upper limit of normal in samples stored 10 years

This tHcy method showed the ability to distinguish between normal and affected patients with different forms of homocystinuria. However, using this method as a second-tier test has limitations. Using elevated Met as a primary marker in routine NBS for classical homcystinuria will miss patients with other causes of homocystinuria, who have normal or low levels of Met. Two patients diagnosed with cblG and MTHFR were missed through the NBS programme with normal values of Met and C3-carnitine (Table 1). Both these patients had borderline low ratios of methionine:phenylalanine (0.20 and 0.21) which could have triggered further testing if ratios were included as a secondary marker, depending on the cutoff used. These results agree with previous reports, where the NBS results of Met were normal or low, with low methionine:phenylal-

anine ratios in the remethylation disorders of homocysteine (Tortorelli et al. 2010; Turgeon et al. 2010).

In summary, we have developed a method for the reliable quantitative analysis of tHcy using the initial DBS of NBS samples, and have demonstrated the utility of this method as a second-tier test for NBS programmes. This method clearly distinguished between normal and affected infants in stored DBSs from 19 patients previously diagnosed as CBS deficient (six patients), cblC deficient (ten patients), cblG deficient (two patients), and one patient with MTHFR deficiency.

All cases were or would have been detected using a second-tier tHcy assay triggered by increased methionine, or increased C3 or C3/C2 ratio, and/or low methionine:phenylalanine ratio on initial NBS.

 Springer

Competing Interests

Nil.

Funding

This research was funded by NSW Biochemical Genetics Service.

Ethical Approval

This research was carried out within the guidelines of The Children's Hospital at Westmead Research Ethics Committee.

Guarantor

K. Carpenter.

Contributorship

A. Alodaib conducted the method development and wrote the first draft of the manuscript. K. Carpenter supervised the project. V. Wiley and T. Wotton provided newborn screening samples and results. J. Christodoulou and B. Wilcken directed the research. All authors reviewed and edited the manuscript and approved the final version of the manuscript.

References

Bowron A, Barton A, Scott J, Stansbie D (2005) Blood spot homocysteine: a feasibility and stability study. Clin Chem 51:257–258

Digest I (2007) Hyperhomocysteinemia and cobalamin disorders. Mol Genet Metab 90:113–121

Fowler B (1997) Disorders of homocysteine metabolism. J Inherit Metab Dis 20:270–285

Gan-Schreier H, Kebbewar M, Fang-Hoffmann J et al (2009) Newborn population screening for classic homocystinuria by determination of total homocysteine from guthrie cards. J Pediatr 156:427–432

Gempel K, Gerbitz K-D, Casetta B, Bauer MF (2000) Rapid determination of total homocysteine in blood spots by liquid chromatography-electrospray ionization-tandem mass spectrometry. Clin Chem 46:122–123

Magera MJ, Lacey JM, Casetta B, Rinaldo P (1999) Method for the determination of total homocysteine in plasma and urine by stable isotope dilution and electrospray tandem mass spectrometry. Clin Chem 45:1517–1522

Matern D, Tortorelli S, Oglesbee D, Gavrilov D, Rinaldo P (2007) Reduction of the false-positive rate in newborn screening by implementation of MS/MS-based second-tier tests: the Mayo Clinic experience (2004–2007). J Inherit Metab Dis 30:585–592

McCann SJ, Gillingwater S, Keevil BG, Cooper DP, Morris MR (2003) Measurement of total homocysteine in plasma and blood spots using liquid chromatography-tandem mass spectrometry: comparison with the plasma Abbott IMx method. Ann Clin Biochem 40:161–165

Refsum H, Fredriksen Å, Meyer K, Ueland PM, Kase BF (2004) Birth prevalence of homocystinuria. J Pediatr 144:830–832

Schiff M, Benoist J-F, Tilea B, Royer N, Giraudier S, Ogier de Baulny H (2011) Isolated remethylation disorders: do our treatments benefit patients? J Inherit Metab Dis 34:137–45

Tortorelli S, Turgeon CT, Lim JS et al (2010) Two-tier approach to the newborn screening of methylenetetrahydrofolate reductase deficiency and other remethylation disorders with tandem mass spectrometry. J Pediatr 157:271–275

Turgeon CT, Magera MJ, Cuthbert CD et al (2010) Determination of total homocysteine, methylmalonic acid, and 2-methylcitric acid in dried blood spots by tandem mass spectrometry. Clin Chem 56:1686–1695

Whitehead VM (2006) Acquired and inherited disorders of cobalamin and folate in children. Br J Haematol 134:125–136

Wilcken B, Wiley V, Hammond J, Carpenter K (2003) Screening newborns for inborn errors of metabolism by tandem mass spectrometry. N Engl J Med 348:2304–2312

Yap S, Naughten E (1998) Homocystinuria due to cystathionine β-synthase deficiency in Ireland: 25 years' experience of a newborn screened and treated population with reference to clinical outcome and biochemical control. J Inherit Metab Dis 21:738–747

JIMD Reports
DOI 10.1007/8904_2011_110

CASE REPORT

Galactokinase Deficiency in a Patient with Congenital Hyperinsulinism

Mashbat Bayarchimeg · Dunia Ismail · Amanda Lam ·
Derek Burk · Jeremy Kirk · Wolfgang Hogler ·
Sarah E Flanagan · Sian Ellard · Khalid Hussain

Received: 5 August 2011 / Revised: 14 September 2011 / Accepted: 25 October 2011 / Published online: 13 December 2011
© SSIEM and Springer-Verlag Berlin Heidelberg 2011

Abstract *Background*: Galactokinase catalyses the first committed step in galactose metabolism, the conversion of galactose to galactose-1-phosphate. Galactokinase deficiency is an extremely rare form of galactosaemia, and the most frequent complication reported is cataracts. Congenital hyperinsulinism (CHI) is a cause of severe hypoglycaemia in the newborn period. Galactosaemia has not previously been reported in a neonate with concomitant CHI.

Aims: To report the first case of a patient with CHI and galactokinase deficiency, and to describe the diagnostic pitfalls with bedside blood glucose testing in a neonate with combined galactokinase deficiency and CHI.

Communicated by: Alberto B Burlina.

Competing interests: None declared.

M. Bayarchimeg · D. Ismail · K. Hussain
Department of Endocrinology, Great Ormond Street Hospital for
Children NHS Trust, London, UK

M. Bayarchimeg · D. Ismail · A. Lam · D. Burk · K. Hussain (✉)
Clinical and Molecular Genetics Unit, Developmental
Endocrinology Research Group, The Institute of Child Health,
University College London, 30 Guilford Street,
London WC1N 1EH, UK
e-mail: K.Hussain@ich.ucl.ac.uk

A. Lam · D. Burk
Department of Biochemistry, Great Ormond Street Hospital for
Children NHS Trust, London, UK

J. Kirk · W. Hogler
Department of Endocrinology and Diabetes, Birmingham
Children's Hospital, Birmingham, UK

S. E. Flanagan · S. Ellard
Institute of Biomedical and Clinical Science,
Peninsula Medical School, University of Exeter, Barrack Road,
Exeter, UK

Patients/methods: A 3-day-old baby girl from consanguineous parents presented with poor feeding, irritability and seizures. Capillary blood glucose testing using bedside test strips and glucometer showed a glucose level of 18 mmol/L, but the actual laboratory blood glucose level was only 1.8 mmol/L. After discontinuation of oral feeding (stopping provision of dietary galactose), the bedside capillary blood glucose correlated with laboratory glucose concentrations. *Results*: Biochemically the patient had CHI (blood glucose level 2.3 mmol/L with simultaneous serum insulin level of 30 mU/L) and galactokinase deficiency (elevated serum galactose level 0.62 μmol/L). Homozygous loss of function mutations in *ABCC8* and *GALK1* were found, which explained the patient's CHI and galactokinase deficiency, respectively.

Conclusion: This is the first reported case of CHI and galactokinase deficiency occurring in the same patient. Severe hypoglycaemia in neonates with CHI may go undetected with bedside blood glucose meters in patients with galactokinase deficiency.

Introduction

Galactosaemia describes a group of diseases characterised by abnormalities in galactose metabolism. Galactokinase deficiency (OMIM 230200) is a rare type of galactosaemia and causes congenital cataracts during infancy (Hennermann et al. 2011). The enzyme galactokinase catalyses the first committed step in the metabolism of galactose by phosphorylating the galactose at the first carbon. Galactokinase is encoded by the *GALK1* gene which is localised on chromosome 17q24 (Stambolian et al. 1995).

Congenital hyperinsulinism (CHI) is a major cause of severe hypoglycaemia in the newborn period. Delay in the

diagnosis and treatment of the hypoglycaemia is the major reason for the increased risk of brain damage observed in these patients. The most common cause of severe CHI which is medically unresponsive involves defects in the genes regulating the function of the pancreatic ATP sensitive potassium channel (K_{ATP} channel) (Thomas et al. 1995, 1996). Mutations in *ABCC8* are the most common cause of severe medically unresponsive CHI.

Blood glucose test strips in conjunction with glucose meters are used widely to measure capillary blood glucose levels in patients with hypoglycaemia and hyperglycaemia (Meex et al. 2006). The advantages of using these glucose meters with the test strips are that they are readily available, they give immediate results and a minimum amount of blood is required. Despite these advantages, the results can be affected by numerous factors. For example, severe dehydration, hypotension and high haematocrit ($>55\%$) may cause an underestimation of blood glucose concentration (Barreau and Buttery 1987; Atkin et al. 1991). Falsely elevated blood glucose levels may be caused by lipaemic blood, low haematocrit ($<25\%$) and chemicals in the blood that are measured by the meter.

We report the first case of combined CHI and galactokinase deficiency in a human. This baby girl presented with fitting and routine glucose measured by bedside test strips and a glucose meter showed a blood glucose level of 18 mmol/L, but the actual laboratory blood glucose (measured using a glucose oxidase method) level was only 1.8 mmol/L. The falsely elevated blood glucose was due to the increased concentration of galactose which the meter was unable to discriminate from blood glucose.

Case History

The patient was born at term with a birth weight of 4.2 kg to consanguineous parents following a normal vaginal delivery. There was no history of gestational diabetes mellitus, and she was discharged home on day 2 of life with no concerns. On day 3, however, she had generalised seizures (eye rolling, sweatiness and shaking limbs) requiring admission to hospital. Routine blood glucose monitoring (whilst on enteral feeds and intravenous dextrose) showed a marked discrepancy between the bedside glucometer (Roche Accu-Chek Advantage II meter; Roche Diagnostics Limited, Lewes, East Sussex, UK) blood glucose reading and the laboratory blood glucose levels (see Table 1) as measured by the glucose oxidase method. This discrepancy disappeared when the feeds were stopped and the patient maintained on intravenous glucose infusion.

Once the feeds were stopped, she required up to 22 mg/ kg/min of intravenous dextrose to maintain normoglycae-

Table 1 Discrepancy between the blood glucose levels measured by the bedside using a glucometer with test strips and laboratory blood glucose

Bedside blood glucose reading (mmol/L)	Laboratory blood glucose (mmol/L)	Milk feeds
18	2.9	Yes
12	3.1	Yes
15	2.2	Yes
5.6	5.8	No
4.3	4.6	No
4.7	4.9	No

mia, and further investigations confirmed hyperinsulinaemic hypoglycaemia (laboratory blood glucose 2.3 mmol/L with a simultaneous serum insulin of 30 mU/L, and undetectable serum fatty acid and ketone bodies). Her hypoglycaemia failed to respond to maximal dose of medical therapy with diazoxide (20 mg/kg/day) and octreotide (35 µg/kg/day), thus requiring a near total pancreatectomy. Histology of the resected pancreas confirmed changes typical of diffuse CHI. Ophthalmology examination showed no cataracts but pseudotumour cerebri was noted. She was then managed on a galactose-free diet. At the age of 2 years, this patient has global developmental delay, microcephaly and generalised epilepsy.

Methods

Genetic Studies for *ABCC8/KCNJ11* Mutations

Genomic DNA was extracted from peripheral leukocytes using standard procedures. The *KCNJ11* and *ABCC8* genes were amplified and the products sequenced as previously described (Flanagan et al. 2007). The sequences were compared to the published sequences (NM_000525 and NM_000352.2) using Mutation Surveyor 3.24 software (SoftGenetics, PA, USA). Mutation testing was undertaken in parental samples to confirm their carrier status.

Genetic Studies for Galactokinase

Genomic DNA was extracted from peripheral leucocytes using standard procedures and the GALK1 was amplified (primers available on request). Standard PCR conditions were used with the addition of 5 mM betaine and 5% DMSO. PCR products were purified using microClean (Web Scientific) and then sequenced using the ABI Big Dye Terminator v1.1 Cycle Sequencing Kit (Applied Biosystems) with standard conditions and then run on the ABI 3730 DNA analyser. Data were analysed using Sequencher 4.9 (Gene Codes Corporation, Ann Arbor, MI, USA).

 Springer

Enzymology

Galactokinase Assay

Galactokinase activity was assayed in washed lysed erythrocytes as described by Ng et al. (1965). Samples were incubated with 14°C-galactose (Amersham, Bucks, UK) in the presence of ATP and magnesium for 30 min at 37°C. The 14°C-galactose-1-phosphate product, radiolabelled reaction intermediates and unreacted 14°C-galactose were separated by descending paper chromatography using DE81 ion exchange paper (Whatman, Kent, UK). Non-specific radiolabelled contaminants were corrected with a sample blank in which the sample was added immediately before the reaction was terminated. Results were expressed as micromols of 14°C-galactose-1-phosphate generated per hour per gram of haemoglobin (μmol/h/g Hb) following correction of non-specific radiolabelled contaminants and radiolabelled reaction intermediates.

Galactose-1-Phosphate Uridyl Transferase Assay

As galactokinase is a labile enzyme, a second enzyme galactose-1-phosphate uridyl transferase was assayed in washed lysed erythrocytes using the method described by Beutler (Beutler and Baluda 1966). Results are expressed as the amount of uridine diphosphoglucose consumed per hour per gram of haemoglobin (μmol/h/g Hb).

Results

Genetic Studies for CHI

Sequence analysis identified a previously reported missense mutation, E128K (c.382 G>A; p.Glu128Lys) in exon 3 of the *ABCC8* gene (Fig. 1) (Yan et al. 2007). Mutation testing confirmed that the unaffected parents were heterozygous for the mutation.

Galactokinase

All the coding regions of *GALK1* gene from the affected patient were sequenced to investigate the underlying genetic causes of the galactokinase-deficient phenotype. We detected a homozygous R256W (c.766 C>T; p.Arg256Trp) missense mutation in exon 5 of the *GALK1* gene (Fig. 2) which encodes the galactokinase protein and which has been previously identified in a different patient and shown to substantially reduce the enzyme activity (Asada et al. 1999). The mutation occurred at CpG dinucleotides but outside of any conserved regions of the gene. Asada et al. (1999) has shown through COS cell

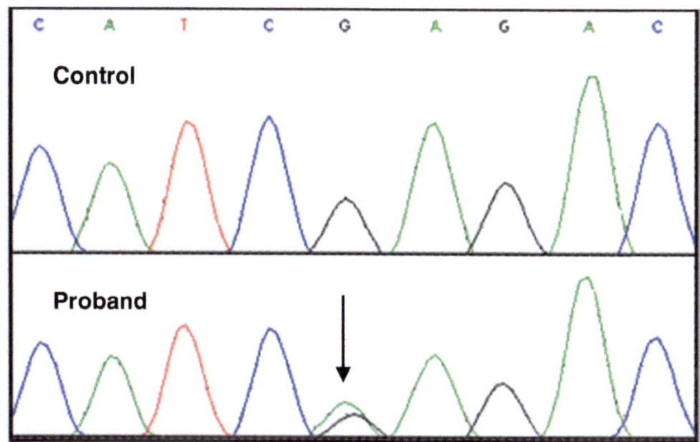

Fig. 1 Electropherograms showing the homozygous *ABCC8* mutation and a normal control. A *black arrow* points to the c.382 G>A mutation which results in the substitution of glutamic acid (GAG) by lysine (AAG) at residue 128 (E128K)

expression analysis that this particular missense mutation completely abolishes the activity of galactokinase.

Enzymology

Galactokinase activity in the patient was undetectable and was significantly lower than a simultaneously assayed control erythrocyte sample (2.1 μmol/h/g Hb, reference range 1.0–3.6). The galactose-1-phosphate uridyl transferase activity in the patient was within the unaffected range (33.7 μmol/h/g Hb, reference range 18.0–40.0). The serum level of galactose at the time of diagnosis was 0.62 μmol/L (normal for non-galactosaemic patient <0.1).

Discussion

This is the first patient to be reported with CHI and galactokinase deficiency. The loss of function mutation in the *ABCC8* gene leads to severe hyperinsulinaemic hypoglycaemia, whereas the loss of function mutation in the *GALK1* gene leads to a substantial reduction in the activity of the enzyme galactokinase resulting in the accumulation of galactose in the blood. The functional consequences of the *GALK1* mutation observed in our patient have been previously studied by Asada et al. (1999). This particular *GALK1* mutation completely abolishes the activity of galactokinase and this was confirmed biochemically in our patient.

Hand-held (bedside) blood glucose monitors in conjunction with glucose test strips are widely used in the care of patients with hypoglycaemia (neonatal hypoglycaemia) and hyperglycaemia (diabetes mellitus). Despite the potential advantages, some of these capillary glucose monitors fail to discriminate between blood glucose and other substrates

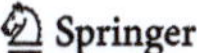

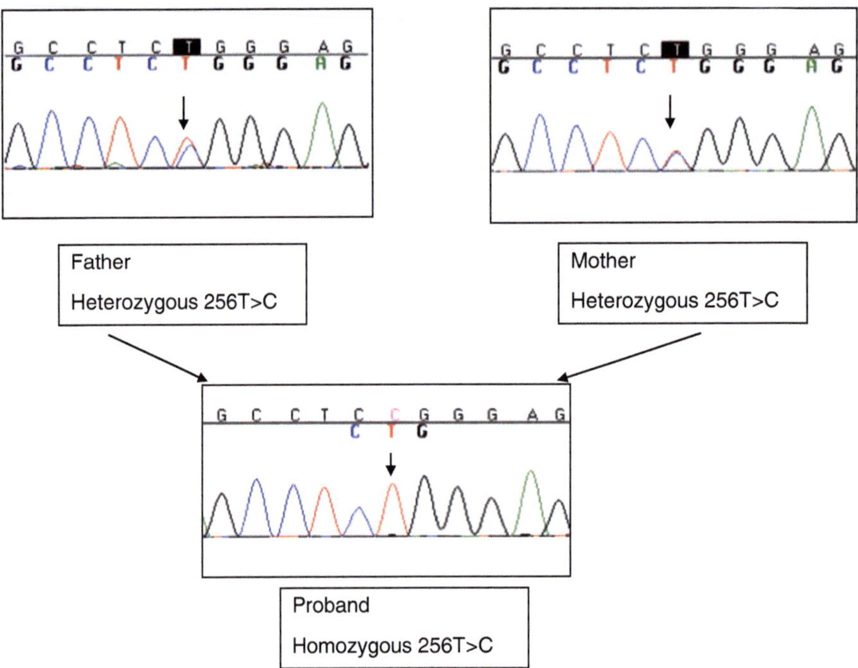

Fig. 2 Electropherograms showing the homozygous *GALK1* mutation in the proband and the heterozygote status of the parents. The *black arrow* points to the homozygous R256W (c.766 C>T; p.Arg256Trp) missense mutation in exon 5 of the *GALK1* gene

such as galactose (Hyde and Betts 2006; Nelson 2008; Newman et al. 2002). The Roche Accu-Chek Advantage II meter used in this patient will detect blood galactose if the galactose concentration is >0.56 mmol/L (Newman et al. 2002). These test strips and Roche Accu-Chek Advantage II meter use the glucose dehydrogenase (GDH) and pyrroloquinolinequinone (PQQ) enzyme system for measuring blood glucose, and this is known to detect galactose as well as glucose. In contrast, the standard laboratory methods used to measure blood glucose levels are based on the glucose oxidase or hexokinase methods.

Galactokinase (GALK) is the first enzyme in the Leloir pathway, converting galactose into galactose-1-phosphate (Gal-1-P). The deficiency of galactokinase leads to an accumulation of galactose in the blood which then interferes with the capillary blood glucose monitoring. In the case of the Roche Accu-Chek Advantage II meter, the increased level of galactose leads to a falsely elevated measurement for blood glucose. There have been several case reports of newborns with classical galactosaemia (due to galactose-1-phosphate uridyl transferase deficiency) who presented with falsely elevated blood glucose levels when measured by the bedside using a Roche Accu-Chek Advantage II meter (Hyde and Betts 2006; Nelson 2008; Newman et al. 2002). Hyde and Betts (2006) reported a 5-week-old galactosaemic baby with failure to thrive and again falsely elevated blood glucose levels. Newman et al. (2002) reported a premature newborn with galactosaemia who had falsely elevated blood glucose levels.

Whilst galactokinase deficiency is rare in comparison to galactosaemia, the gene frequency, however, can vary widely with an east-to-west gradient across Europe (from 1:1,000,000 to 1:52,000). A high incidence of galactokinase deficiency is found among Roma, an endogamous Gypsy population originating from Eastern Europe (16). The high incidence is attributable to a founder effect, as demonstrated by the segregation of a single nucleotide mutation (P28T) which is present in about 5% of the Roma population (Kalaydjieva et al. 1999).

In Galactokinase deficiency, cataracts and pseudotumour cerebri appear to be the major complications, and the outcome for patients with galactokinase deficiency is thought to be much better than for patients with classical galactosaemia (Hennermann et al. 2011). In contrast, in classical galactosaemia long-term follow-up studies of patients have shown that, in spite of a severely galactose-restricted diet, most patients develop abnormalities such as disturbed mental and/or motor development and females develop hypergonadotropic hypogonadism (Bosch et al. 2002). In classical galactosaemia there is also impairment of speech, resulting from disruption in motor planning and programming or motor execution (Potter 2011). In our patient the global developmental delay, microcephaly and generalised epilepsy most likely reflect the combination of hyperinsulinaemic hypoglycaemia and galactokinase deficiency.

In summary, we present a previously unreported combination of a neonate with CHI and galactokinase deficiency. The homozygous *ABCC8* gene mutation led to severe hypoglycaemia, and the homozygous *GALK1* mutation led to the accumulation of galactose. The elevated serum galactose level resulted in falsely high capillary blood glucose level when measured by a bedside glucometer, and obscured the prevailing hypoglycaemia caused by CHI. Severe hypoglycaemia in neonates with CHI may go undetected with bedside capillary blood glucose meters in the presence of galactokinase deficiency.

Disclosure Summary

No conflict of interest for any authors.

Precis

This is the first report of an infant with combined galactokinase deficiency and congenital hyperinsulinism. Galactokinase deficiency led to falsely elevated blood glucose levels in the presence of severe hypoglycaemia caused by congenital hyperinsulinism.

References

Asada M, Okano Y, Imamura T, Suyama I, Hase Y, Isshiki G (1999) Molecular characterization of galactokinase deficiency in Japanese patients. J Hum Genet 44(6):377–382

Atkin SH, Dasmahapatra A, Jaker MA, Chorost MI, Reddy S (1991) Fingerstick glucose determination in shock. Ann Intern Med 114 (12):1020–1024

Barreau PB, Buttery JE (1987) The effect of the haematocrit value on the determination of glucose levels by reagent-strip methods. Med J Aust 147(6):286–288

Beutler E, Baluda MC (1966) Improved method for measuring galactose-1-phosphate uridyl transferase activity of erythrocytes. Clin Chim Acta 13:369–379

Bosch AM, Bakker HD, van Gennip AH, van Kempen JV, Wanders RJ, Wijburg FA (2002) Clinical features of galactokinase deficiency: a review of the literature. J Inherit Metab Dis 25(8):629–634

Flanagan SE, Patch AM, Mackay DJ, Edghill EL, Gloyn AL, Robinson D, Shield JP et al (2007) Mutations in ATP-sensitive K + channel genes cause transient neonatal diabetes and permanent diabetes in childhood or adulthood. Diabetes 56 (7):1930–1937

Hennermann JB, Schadewaldt P, Vetter B, Shin YS, Mönch E, Klein J (2011) Features and outcome of galactokinase deficiency in children diagnosed by newborn screening. J Inherit Metab Dis 34 (2):399–407

Hyde P, Betts P (2006) Galactosaemia presenting as bedside hyperglycaemia. J Paediatr Child Health 42(10):659

Kalaydjieva L, Perez-Lezaun A, Angelicheva D, Onengut S, Dye D, Bosshard NU, Jordanova A et al (1999) A founder mutation in the GK1 gene is responsible for galactokinase deficiency in Roma (Gypsies). Am J Hum Genet 65:1299–1307

Meex C, Poncin J, Chapelle JP, Cavalier E (2006) Analytical validation of the new plasma calibrated Accu-Chek Test Strips (Roche Diagnostics). Clin Chem Lab Med 44(11):1376–1378

Nelson JR (2008) False bedside hyperglycaemia revealing galactosaemia in a premature twin female baby. Arch Pediatr 15 (8):1359–1360

Newman JD, Ramsden CA, Balazs ND (2002) Monitoring neonatal hypoglycemia with the Accu-chek advantage II glucose meter: the cautionary tale of galactosemia. Clin Chem 48(11):2071

Ng WG, Donnell GN, Bergren WR (1965) Galactokinase activity in human erythrocytes of individuals at different ages. J Lab Clin Med 66:115–121

Potter NL (2011) Voice disorders in children with classic galactosaemia. J Inherit Metab Dis 34(2):377–385

Stambolian D, Ai Y, Sidjanin D, Nesburn K, Sathe G, Rosenberg M, Bergsma DJ (1995) Cloning of the galactokinase cDNA and identification of mutations in two families with cataracts. Nat Genet 10:307–312

Thomas P, Ye Y, Lightner E (1996) Mutation of the pancreatic islet inward rectifier Kir6.2 also leads to familial persistent hyperinsulinemic hypoglycemia of infancy. Hum Mol Genet 5 (11):1809–1812

Thomas PM, Cote GJ, Wohllk N, Haddad B, Mathew PM, Rabl W, Aguilar-Bryan L et al (1995) Mutations in the sulfonylurea receptor gene in familial persistent hyperinsulinemic hypoglycemia of infancy. Science 268:426–429

Yan FF, Lin YW, MacMullen C, Ganguly A, Stanley CA, Shyng SL (2007) Congenital hyperinsulinism associated ABCC8 mutations that cause defective trafficking of ATP-sensitive K + channels: identification and rescue. Diabetes 56(9):2339–2348

JIMD Reports
DOI 10.1007/8904_2011_111

Heart Failure Due to Severe Hypertrophic Cardiomyopathy Reversed by Low Calorie, High Protein Dietary Adjustments in a Glycogen Storage Disease Type IIIa Patient

Christiaan P. Sentner · Kadir Caliskan · Wim B. Vletter · G. Peter A. Smit

Received: 23 August 2011 / Revised: 30 September 2011 / Accepted: 25 October 2011 / Published online: 13 December 2011
© SSIEM and Springer-Verlag Berlin Heidelberg 2011

Abstract In glycogen storage disease type III (GSD III), deficiency of the debranching enzyme causes storage of an intermediate glycogen molecule (limit dextrin) in the affected tissues. In subtype IIIa hepatic tissue, skeletal- and cardiac muscle tissue is affected, while in subtype IIIb only hepatic tissue is affected. Cardiac storage of limit dextrin causes a form of cardiomyopathy, which resembles primary hypertrophic cardiomyopathy on cardiac ultrasound. We present a 32-year-old GSD IIIa patient with severe left ventricular hypertrophy (LVH) first diagnosed at the age of 8 years. LVH remained stable and symptomless until the patient presented at age 25 years with increasing dyspnea, fatigue, obesity, and NYHA (New York Heart Association) functional classification two out of four. Dyspnea, fatigue, and obesity progressed, and at age 28 years she was severely symptomatic with NYHA classification 3+ out of 4. On echocardiogram and electrocardiogram, the LVH had progressed as well. Initially, she was rejected for cardiac transplantation because of severe obesity. Therefore, a 900 cal, high protein diet providing 37% of total energy was prescribed during 4 months on which 10 kg weight loss was achieved. However, her symptoms as well as the electrocardiographic and echocardiographic LVH indices had improved dramatically – ultimately deferring cardiac transplantation. Thereafter, the caloric intake was increased to 1,370 cal per day, and the high protein intake was continued providing 43% of total energy. After 3 years of follow-up, the patient remains satisfied with reasonable exercise tolerance and minor symptoms in daily life.

Introduction

Glycogen storage disease type III (GSD III) is an autosomal recessive disorder in which a mutation in the *AGL* gene causes deficiency of the debranching enzyme (DE). The DE consists of two active centers, which catalyze the last step in the conversion of glycogen to glucose (Smit et al. 2006). The absence of DE activity in GSD III patients causes storage of an intermediate form of glycogen, limit dextrin (LD) (Chen 2001). Eighty-five percent of the GSD III patients have subtype IIIa in which DE is deficient in muscle and liver tissue. Fifteen percent of the patients have subtype IIIb in which DE is deficient in the liver (Shen et al. 1996). In neonates and infants, the main features are hepatomegaly, keto-hypoglycemic episodes after short periods of fasting, and hyperlipidemia. Poorly treated neonates and children have developmental delay, growth retardation, and delayed puberty. Proximal and distal myopathy presents in adult GSD IIIa patients, which may be enforced by the development of peripheral neuropathy (Wolfsdorf and Weinstein 2003). Cardiomyopathy is a frequent complication in GSD IIIa since LD can also store in myocardial cells and between bundles of myofilaments (Moses et al. 1989; Smit et al. 1990; Labrune et al. 1991;

Communicated by: Alberto B Burlina.

Competing interests: None declared.

C.P. Sentner (✉) · G.P.A. Smit
Department of Metabolic Diseases, Beatrix Children's Hospital, University Medical Centre Groningen, Hanzeplein 1, P.O. Box 30.001, 9700 RB, Groningen, The Netherlands
e-mail: c.p.sentner@bkk.umcg.nl

K. Caliskan · W.B. Vletter
Department of Cardiology, Erasmus University Medical Centre, Rotterdam, The Netherlands

Coleman et al. 1992; Carvalho et al. 1993; Talente et al. 1994). This causes a form of cardiomyopathy that echocardiographically resembles primary hypertrophic cardiomyopathy due to sarcomere gene mutations, but shows a different response to exercise testing, 24-h electrocardiographic monitoring and thallium-201 myocardial scintigraphy (Lee et al. 1997; Akazawa et al. 1997; Olson et al. 1984). The clinical significance and long-term consequences of GSD IIIa-related cardiomyopathy are unclear due to a lack of data and experience.

The aim of the dietary treatment of GSD III is to divide the carbohydrate intake throughout the day to maintain normoglycemia by taking frequent meals and regular cornstarch doses (Gremse et al. 1990). Protein supplementation is necessary as it serves as a substrate for gluconeogenesis during fasting conditions and improves myopathy and growth failure (Slonim et al. 1982, 1984; Kiechl et al. 1999). However, there is no consensus between centers on the usage of a high protein diet or the amount of cornstarch that should be provided.

In this case report, we present a GSD IIIa patient with severely symptomatic hypertrophic cardiomyopathy, which was reversed after initiating a low calorie, high protein diet to achieve weight loss for a cardiac transplantation preparation program. Subsequently, her cardiomyopathy-related symptoms and signs improved dramatically and cardiac transplantation could be deferred.

Case Report

The patient reported is a 32-year-old Turkish female born to consanguineous parents after an uncomplicated pregnancy and birth. She was diagnosed with glycogen storage disease in Turkey at a young age, but subtyping for GSD III was only initiated at the age of 8 years after she moved to the Netherlands. Enzymatic measurements confirmed absent DE activity in muscle- and liver tissue, confirming GSD IIIa. Genetic mutation analysis revealed a pathologic homozygote four basepair deletion in exon 7 of the *AGL* gene (GeneBank genomic reference sequence NW_012865) c.753_756delCAGA causing a frameshift (Lucchiari et al. 2006). Upon clinical evaluation, the main findings were hepatomegaly with elevated aspartate transaminase (203 U/L), alanine transferase (253 U/L), triglycerides (1.7 mmol/L), and creatine kinase values (2,112 U/L). A grade III out of VI systolic cardiac murmur was heard in the 4th left intercostal space. Further cardiac evaluation revealed hypertrophic cardiomyopathy with concentric left ventricular hypertrophy (LVH). On the ECG, no rhythm- or conduction disturbances were seen. Following diagnosis, dietary treatment was initiated with protein-enriched frequent meals during the day and one late night meal. She responded well to the treatment and no hypoglycaemic episodes requiring hospitalization have occurred since. During puberty, the liver size normalized, but the transaminase values remained elevated. Echocardiographically the LVH remained stable and symptomless; therefore, no further cardiologic follow-up was deemed necessary.

At age 25 years, the patient presented in the outpatient-ward with increasing dyspnea, fatigue, obesity, and functional classification according to the New York Heart Association (NYHA) two out of four (Table 1). On physical examination, her heart rate was 69 bpm, blood pressure 100/60 mmHg, with mild jugular venous pressure elevation. Her electrocardiogram showed increased QRS-voltage and -duration with negative T-waves comparable with severe LVH (Fig. 1). The Sokolow-Lyon-, Cornell Voltage- and Romhilt-Estes electrocardiographic indices for LVH were all positive (Fig. 1). The echocardiogram showed severe concentric LVH with intraventricular septum (IVS) thickness of 22 mm, and a posterior wall (PW) thickness of 18 mm with normal dimensions (Table 1). She was treated with low-dose furosemide, fluid restriction, and her diet was adjusted to provide extra protein during the day. Also, a late night feed combined with cornstarch was added to ensure normoglycemia during the night. In the following years, the symptoms of dyspnea and fatigue slowly progressed, and she gained weight. At age 28 years, she was severely symptomatic with a NYHA functional class of 3+ out of 4. The patient's BMI increased to 32.2 because she avoided physical exercise as this provoked palpitations and chest pain. The electrocardiographic and echocardiographic LVH indices worsened accordingly (IVS thickness 32 mm, PW thickness 25 mm). At that time low-dose perindopril and carvedilol were added to the furosemide, and a prophylactic implantable cardioverter defibrillator was placed due to an increased risk of sudden cardiac death.

Due to the worsening situation, the patient was evaluated for heart transplantation (HTX) at the age of 30 years. Therefore, a pre-HTX program was set up consisting of: (1) evaluation of the status of liver and skeletal muscle, (2) a new dietary regimen to reduce weight by 10 kg, (3) perioperative advice regarding the management of GSD III during the HTX. The new dietary regimen consisted of 24-h protein-enriched naso-gastric drip feeding containing 900 cal per day, with protein providing 37%, carbohydrates 61%, and lipids 2% providing of total energy. After following the new dietary regimen for 4 months, her BMI decreased to 27.7, along with a significant improvement of her complaints of constant fatigue and exercise intolerance. The NYHA classification decreased accordingly to two out of four. On cardiac ultrasound, concentric LVH was still present but the PW and IVS thickness had

Table 1 Clinical symptoms, laboratory and echocardiographic results over time

Age (years)	25	26	27	28	29	30[a]	31[a]	32[a]
BMI[b]	29.7	29.7	30.9	30.3	32.2	**30.5**	**27.7**	**27.8**
Dyspnea[c]	+	+	+	++	++	−	−	−
Chest Pain[c]	±	±	±	+	++	−	−	−
Fatigue[c]	++	+	++	++	++	+	+	+
Palpitations[c]	−	−	−	±	+	−	−	−
NYHA classification[d]	2	2	2	3+	3+	**2**	**2**	**2**
Creatin kinase (U/L)	578	1,368	1,537	3,662	2,712	**849**	**1,449**	**1,400**
Echocardiographic measurements (mm)								
IVS[e]	22	27	32	28	32	**29**	**22**	**21**
PW[f]	18	23	24	29	25	**25**	**25**	**25**
LA[g]	34	31	33	41	42	**47**	**47**	**48**
LVED[h]	49	45	46	42	43	**43**	**44**	**49**
LVES[i]	30	34	29	27	30	**28**	**31**	**33**

[a] Bold numbers indicate data after the introduction of a high-protein, low calorie diet at the age of 30 years

[b] BMI indicates body mass index

[c] − Indicates no symptoms; ± minor symptoms; + indicates intermittent symptoms; ++ indicates daily symptoms

[d] NYHA classification indicates functional classification for heart failure by the New York Heart Association

[e] IVS indicates interventricular septum

[f] PW indicates left ventricular posterior wall

[g] LA indicates left atrial dimensions

[h] LVED indicates left ventricular end-diastolic dimensions

[i] LVES indicates left ventricular end-systolic dimensions

decreased to 25 mm and 24 mm, respectively, along with the electrocardiographic LVH indices. Consequently, the patient was taken off the pre-HTX program, and cardiac transplantation was deferred. The caloric intake was increased to 1,370 cal per day, and the continuous nasogastric drip feeding was stopped. The dietary regimen was switched to seven meals at daytime with 2-h intervals. Five of these meals were drinks, and two were normal meals, which were comparable in composition of energy and protein to the drinks. In the nocturnal period, two gifts of cornstarch were added to reach a 4-h interval between meals. The high protein nature of the diet was maintained, and even increased to provide 43% of total energy per day compared to 37% protein per day during the pre-HTX program. After 3 years of follow-up, she remains satisfied with reasonable exercise tolerance and minor symptoms in daily life.

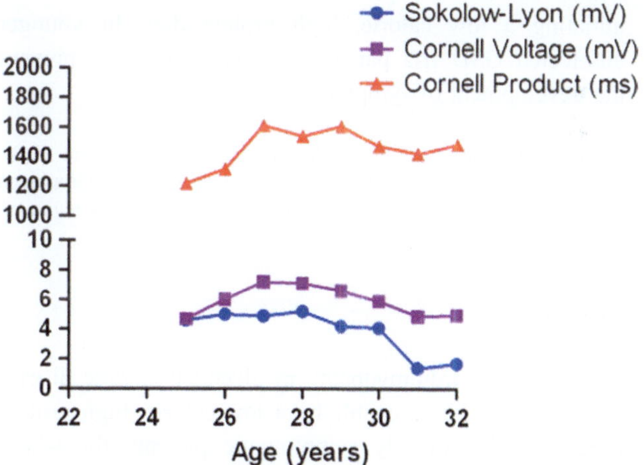

Fig. 1 Twelve-lead electrocardiograpy showing severe left ventricular hypertrophy with secondary repolarization abnormalities at the age of 25, 29, 30, and 32 years old

Discussion

We report a severely symptomatic GSD IIIa patient with cardiomyopathy who improves clinically and objectively on a high protein diet with a limited supply of carbohydrates. To our knowledge, a similar case has been reported recently in a pediatric patient (Valayannopoulos et al. 2011), and once in an adult patient by Dagli et al. in 2009 (Dagli et al. 2009). The latter describes a 22-year-old male with severe GSD III-related cardiomyopathy treated with a high protein diet in which overtreatment with cornstarch was avoided. Their patient improved dramatically with reversal of symptoms and echocardiographic signs of hypertrophic cardiomyopathy.

The physiology of the reversal of extreme LVH in GSD IIIa is not clear. A direct effect, where the limited supplementation of carbohydrate and increased usage of protein in gluconeogenesis reduces the cardiac storage of LD, is feasible. An indirect effect is also feasible where the weight reduction reduces fat disposition in skeletal muscles, improving the condition and possible workload of the skeletal muscles and reducing the workload on the heart (Kelley et al. 1991; Goodpaster et al. 1999). A combination of these effects is also possible.

The majority of the patients seem to remain asymptomatic and the cardiomyopathy seems to be nonprogressive (Lee et al. 1997). Therefore, severe cases such as these emphasize the need for regular cardiac follow-up in patients with GSD IIIa-related cardiomyopathy, along with the benefits of a high protein diet. In our population of GSD III patients, a high protein diet is implemented at a young age and continued through into adulthood. Through these dietary regimens, GSD III patients usually derive 20–30% of total energy from protein, which varies from 4 g of protein per kilogram body weight in children and adolescents to 2 g of protein per kilogram body weight in adult patients. Therefore, renal function is also closely monitored during follow-up of these patients.

This report presents the first case of a GSD IIIa patient in whom cardiac transplantation could be deferred after initiating a low calorie, high protein diet. In younger unaffected GSD IIIa patients, this approach may prevent the development of symptomatic cardiomyopathy.

Acknowledgments We thank Margreet van Rijn, Ph.D. and Els B. Haagsma, M.D., Ph.D. of the University Medical Centre Groningen for providing information on the natural course and dietary regimen.

Synopsis

Hypertrophic cardiomyopathy in glycogen storage disease type IIIa may be reversible by a low calorie, high protein diet, even in severely symptomatic patients for whom cardiac transplantation is being considered.

References

Akazawa H, Kuroda T, Kim S, Mito H, Kojo T, Shimada K (1997) Specific heart muscle disease associated with glycogen storage disease type III: clinical similarity to the dilated phase of hypertrophic cardiomyopathy. Eur Heart J 18:532–533

Carvalho JS, Matthews EE, Leonard JV, Deanfield J (1993) Cardiomyopathy of glycogen storage disease type III. Heart Vessels 8:155–159

Chen YT (2001) Glycogen storage diseases. In: Scriver CR et al (eds) The metabolic and molecular bases of inherited disease. McGraw-Hill Professional, New York, pp 1521–1551

Coleman RA, Winter HS, Wolf B, Gilchrist JM, Chen YT (1992) Glycogen storage disease type III (glycogen debranching enzyme deficiency): correlation of biochemical defects with myopathy and cardiomyopathy. Ann Intern Med 116:896–900

Dagli AI, Zori RT, McCune H, Ivsic T, Maisenbacher MK, Weinstein DA (2009) Reversal of glycogen storage disease type IIIa-related cardiomyopathy with modification of diet. J Inherit Metab Dis Epub Mar 30

Goodpaster BH, Kelley DE, Wing RR, Meier A, Theate FL (1999) Effects of weight loss on insulin sensitivity in obesity – influence of regional adiposity. Diabetes 48:839–847

Gremse DA, Bucuvalas JC, Balisteri WF (1990) Efficacy of cornstarch therapy in type III glycogen-storage disease. Am J Clin Nutr 52:671–674

Kelley DE, Slasky S, Janosky J (1991) Effects of obesity and non-insulin dependent diabetes mellitus. Am J Clin Nutr 54:509–515

Kiechl S, Willeit J, Vogel W, Kohlendorfer U, Poewe W (1999) Reversible severe myopathy of respiratory muscles due to adult-onset type III glycogenosis. Neuromuscul Disord 9:408–410

Labrune P, Huguet P, Odievre M (1991) Cardiomyopathy in glycogen-storage disease type III: clinical and echographic study of 18 patients. Pediatr Cardiol 12:161–163

Lee PJ, Deanfield JE, Burch M, Baig K, McKenna WJ, Leonard JV (1997) Comparison of the functional significance of left ventricular hypertrophy in hypertrophic cardiomyopathy and glycogenosis type III. Am J Cardiol 79:834–838

Lucchiari S, Pagliarani S, Salani S et al (2006) Hepatic and neuromuscular forms of glycogenosis type III: nine mutations in *AGL*. Hum Mutat 27:600–601

Moses SW, Wanderman KL, Myroz A, Frydman M (1989) Cardiac involvement in glycogen storage disease type III. Eur J Pediatr 148:764–766

Olson LJ, Reeder GS, Noller KL, Edwards WD, Howell RR, Michels VV (1984) Cardiac involvement in glycogen storage disease III: morphologic and biochemical characterization with endomyocardial biopsy. Am J Cardiol 53:980–981

Shen J, Bao Y, Liu HM, Lee PJ, Leonard JV, Chen YT (1996) Mutations in exon 3 of the glycogen debranching enzyme gene are associated with glycogen storage disease type III that is differentially expressed in liver and muscle. J Clin Invest 98:352–357

Slonim AE, Weisberg C, Benke P, Evans OB, Burr IM (1982) Reversal of debrancher deficiency myopathy by the use of high-protein nutrition. Ann Neurol 11:420–422

Slonim AE, Coleman RA, Moses SW (1984) Myopathy and growth failure in debrancher enzyme deficiency: improvement with high protein noctural enteral therapy. J Pediatr 105:906–911

Smit GP, Fernandes J, Leonard JV et al (1990) The long-term outcome of patients with glycogen storage diseases. J Inherit Metab Dis 13:411–418

Smit GPA, Rake JP, Akman HO, DiMauro S (2006) The glycogen storage diseases and related disorders. In: Fernandes J, Saudubray J-M, van den Berghe G, Walter JH (eds) Inborn metabolic diseases: diagnosis and treatment. Springer Medizin, Heidelberg, pp 103–116

Talente GM, Coleman RA, Alter C et al (1994) Glycogen storage disease in adults. Ann Intern Med 120:218–226

Valayannopoulos V, Bajolle F, Arnoux JB et al (2011) Successful treatment of severe cardiomyopathy in glycogen storage disease type III with D,L-3-hydroxybutyrate, ketogenic and high protein diet. Pediatr Res 70(6):638–641

Wolfsdorf JI, Weinstein DA (2003) Glycogen storage diseases. Rev Endocr Metab Disord 4:95–102

JIMD Reports
DOI 10.1007/8904_2011_85

Unusual Cardiac "Masses" in a Newborn with Infantile Pompe Disease

Daniel T. Swarr · Beth Kaufman · Mark A. Fogel · Richard Finkel · Jaya Ganesh

Received: 20 April 2011 /Revised: 17 July 2011 /Accepted: 4 August 2011 /Published online: 13 December 2011
© SSIEM and Springer-Verlag Berlin Heidelberg 2011

Abstract Glycogen storage disease type II (OMIM #232300), or Pompe disease, may present in the newborn period with moderate-to-severe biventricular hypertrophy with or without left ventricular outflow tract obstruction that typically leads to death from cardiorespiratory failure in the first year of life. Glycogen deposition tends to be uniform, and is only occasionally accompanied by patchy areas of fibrosis. Here, we present an infant identified with biventricular hypertrophy and cardiac masses by prenatal ultrasound. Postnatal molecular studies did not support the diagnosis of tuberous sclerosis in this case. Additional evaluation for infantile hypertrophic cardiomyopathy confirmed the diagnosis of Pompe disease. We discuss whether the "cardiac masses," which brought this infant to medical attention and facilitated an early diagnosis of Pompe disease, may represent an unusual manifestation of GSD type II or the coincidental occurrence of an unrelated disease process.

Communicated by: Olaf Bodamer.

Competing interests: None declared.

Electronic supplementary material The online version of this chapter (doi:10.1007/8904_2011_85) contains supplementary material, which is available to authorized users.

D.T. Swarr · J. Ganesh (✉)
Department of Pediatrics, Section of Biochemical Genetics, The Children's Hospital of Philadelphia, 34th St & Civic Center Blvd, Philadelphia, PA 19104, USA
e-mail: ganesh@email.chop.edu

B. Kaufman · M.A. Fogel · R. Finkel · J. Ganesh
Department of Pediatrics, University of Pennsylvania School of Medicine, Philadelphia, PA 19104, USA

B. Kaufman · M.A. Fogel
Department of Pediatrics, Division of Cardiology, The Children's Hospital of Philadelphia, Philadelphia, PA 19104, USA

R. Finkel
Department of Pediatrics, Division of Neurology, The Children's Hospital of Philadelphia, Philadelphia, PA 19104, USA

Introduction

Glycogen storage disease type II (OMIM #232300), also known as Pompe disease, is due to a deficiency of the lysosomal enzyme acid-alpha-glucosidase (GAA, E.C. 3.2.1.20). GAA is needed for the hydrolysis of both alpha-1,4- and alpha-1,6-glucosidic linkages within glycogen. Disease severity and age of onset are highly variable, with individuals presenting as early as birth or as late as the seventh decade of life. The classic infantile form is characterized by hypertrophic cardiomyopathy and hypotonia of the skeletal muscle, which results in cardiorespiratory failure and death at an average age of 8.7 months without treatment (Kishnani et al. 2006).

Echocardiography invariably reveals a hypertrophic cardiomyopathy with or without left ventricular outflow tract obstruction, which may progress to a dilated cardiomyopathy in advanced stages of the disease. To our knowledge, discrete masses within the myocardium have never been reported in infantile Pompe disease. We report a 1-month-old male with type II GSD and a discrete cardiac mass detected with routine prenatal ultrasound, which was confirmed with postnatal cardiac imaging. We briefly discuss cardiac imaging in Pompe disease, as well as the possibility that the abnormality may be related to abnormal glycogen storage rather than an incidental finding.

Case Report

The patient first came to medical attention when a routine prenatal ultrasound at 30 weeks gestation identified an intracardiac "mass" and biventricular hypertrophy. Fetal echocardiography confirmed these findings, but no additional prenatal testing was performed before he was delivered at 38 and 0/7 week gestation via emergent cesarean section because of fetal distress. This pregnancy was otherwise uncomplicated and was the second for his 27-year-old mother. The couple's first child is 15 months of age and healthy. No gestational diabetes was reported. His birth weight was 4,560 g (>95th percentile). He did well immediately after birth, and did not exhibit any cardiac, respiratory, or feeding difficulties despite close monitoring in an intensive care setting. No dysmorphic features were noted. Muscle tone was normal limits, and no other abnormal features were noted on neurologic exam. Family medical history was notable for tuberous sclerosis in the patient's father's first cousin. Additional information regarding this individual was unavailable. No intervening family members had been identified as affected, and the father exhibited no symptoms or outward signs of the disorder upon assessment in our clinic. There was no history of consanguinity in the family.

Echocardiography at 27 days of age demonstrated moderate biventricular hypertrophy, and a solitary echogenic mass within the left ventricle, measuring 1 cm × 0.5 cm (see Fig. 1 and supplementary Movie 1). Short-axis images revealed the possibility of an additional mass within the left ventricle adjacent to the papillary muscles, and

subcostal sagittal imaging revealed two additional areas of increased echogenicity within the right ventricular cavity on the septal surface, measuring 7 mm × 4.5 mm and 5 mm × 5 mm. A 12-lead electrocardiogram demonstrated right axis deviation and changes consistent with biventricular hypertrophy, but a normal PR-interval (see Fig. 2).

Cardiac magnetic resonance imaging (MRI) at 1 month of age demonstrated moderate global cardiomegaly with moderate concentric biventricular hypertrophy. Three areas of increased signal intensity on T2- and T1-weighted imaging suggestive of mass-like lesions were visualized in the anteroseptal wall at the mid-short-axis level, the inferoseptal wall near the apex, and within the papillary muscles. The increased signal intensity in the anteroseptal region corresponded to the "mass" visualized by echocardiography. This mass was hyperintense on T1- and T2-weighted imaging, and isointense on steady-state free procession, HASTE, and gradient-echo imaging (see Fig. 3a–c). Applying fat saturation did not affect signal intensity of the mass, and there did not appear to be perfusion to this region on perfusion imaging. On myocardial tagging, the mass did appear to contract. On delayed enhancement imaging, no increased signal intensity was noted.

Additional evaluation was done to exclude other genetic causes of infantile hypertrophic cardiomyopathy, especially those amenable to treatment. The serum creatine kinase was elevated (478 U/L; normal 60–305 U/L). Normal results were noted for levels of plasma amino acids, lactate and pyruvate, total and free carnitine, and acylcarnitine compounds. Urine organic acid quantitation was normal.

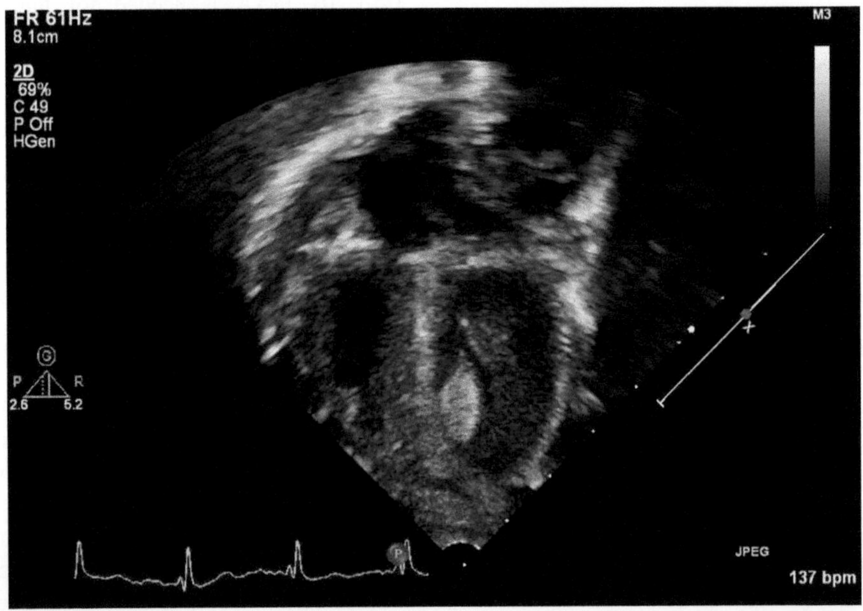

Fig. 1 Echocardiography at 27 days of age demonstrates moderate biventricular hypertrophy and a 1 cm × 0.5 cm discrete echogenic mass within the left ventricle

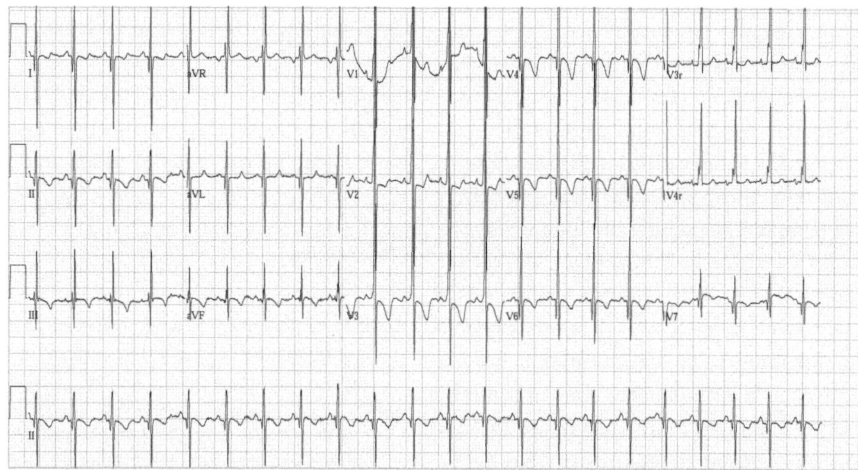

Fig. 2 Electrocardiography demonstrates right axis deviation and changes consistent with biventricular hypertrophy, but a normal PR interval

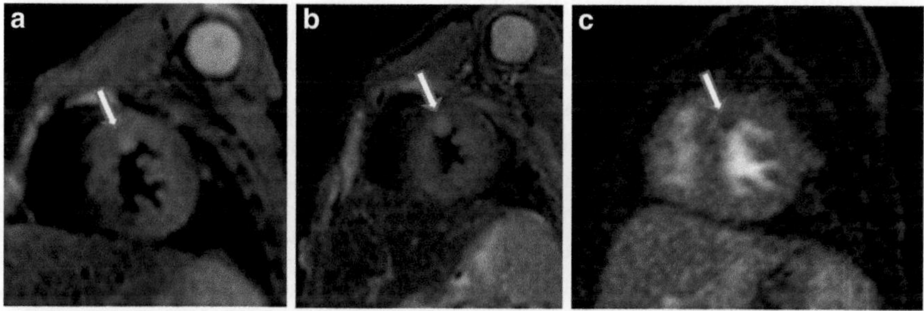

Fig. 3 The mass appears hyperintense on both T1- (**a**) and T2-weighted (**b**) magnetic resonance imaging, and as a "filling defect" on perfusion-weighted imaging (**c**)

GAA activity in whole blood from a blood spot (Duke University metabolic lab; Durham, NC) was 1.8 pmol/punch/hour (control 10–49 pmol/punch/hour), consistent with a diagnosis of infantile Pompe disease. Sequencing of the GAA gene identified two mutations: c.525delT and c.1927 G > A (p.Gly643Arg). Cross-Reactive-Immunological Material (CRIM) testing by western blot analysis of the patient's fibroblasts was positive. Due to the presence of an intracardiac mass and a remote family history of tuberous sclerosis, blood was sent for *TSC1* and *TSC2* gene sequencing and deletion/duplication analysis. MRI of brain was normal, as was analysis of the *TSC1* and *TSC2* genes.

Clinical follow-up at 6 weeks of age demonstrated the interval development of mild hepatomegaly, decreased muscle tone, "hypotonic facies," and brisk reflexes in the lower extremities with bilateral ankle clonus. His cardiac function remained stable, both clinically and by echocardiography.

Discussion

Cardiac imaging in Pompe disease typically reveals uniform hypertrophy due to abnormal glycogen accumulation.

In some cases, there may also be left ventricular outflow tract obstruction due to severe hypertrophy. (Ehlers et al. 1962; Hohn et al. 1965; Rees et al. 1976; Shapir and Roguin 1985) However, discrete or patchy abnormalities within the myocardium are rarely seen, and to our knowledge, discrete cardiac masses have never been reported in an infant with Pompe disease, either by echocardiogram or cardiac MRI.

Only recently has the use of cardiac MRI been described for the assessment of cardiac structure and function in infants with Pompe disease. Barker et al. were able to quantitate biventricular hypertrophy and measure hemodynamics parameters in these infants ($n = 10$). They also identified areas of fibrosis through delayed enhancement with gadolinium (Barker et al. 2010). Myocardial fibrosis, that may be patchy, has been reported in other lysosomal storage disorders, including Fabry disease (De Cobelli et al. 2009). However, it was seen in only one of ten patients in the study of Barker et al., and can be excluded as the cause of the "masses" in our patient based on their overall general appearance and lack of delay in gadolinium enhancement.

The imaging characteristics of this infant's cardiac masses were felt to be most consistent with rhabdomyomas.

This raises the possibility of the concurrence of tuberous sclerosis with infantile Pompe disease. To our knowledge, these two disorders have never been reported in association with one another. Sequencing and duplication/deletion analysis of both *TSC1* and *TSC2*, as well as a chromosomal microarray analysis did now show any deleterious genetic changes. This does not completely exclude the diagnosis, as the estimated sensitivity of combined testing is 70–80% (Au et al. 2007; Dabora et al. 2001; Jones et al. 1999; Sancak et al. 2005). Moreover, somatic mosaicism has been well described in the tuberous sclerosis complex, and if limited to the cardiac tissue in this patient, could account for the multiple cardiac masses with negative DNA sequencing performed on peripheral leukocytes (Verhoef et al. 1999). This infant had no other clinical stigmata of TS, such as ashleaf macules, nor any MRI features.

Alternatively, these cardiac masses may represent uneven and irregular areas of glycogen deposition within the cardiac tissue, simulating the appearance of multiple masses. A third possibility is that these cardiac masses are neither rhabdomyomas, nor abnormal glycogen deposits, but are merely another, incidentally discovered, tumor type. Only cardiac catheterization with biopsy of the mass for histopathologic analysis and *TSC1/TSC2* gene sequencing on tumor tissue would be able to make this distinction, an invasive procedure that is not currently indicated in this infant.

Cardiac masses in the fetus and neonate are rare, and have previously never been reported in association with infantile Pompe disease. In this instance, cardiac masses and biventricular hypertrophy were detected by routine prenatal ultrasound. The case highlights not only a potentially unusual manifestation infantile Pompe, but also the importance of including Pompe disease in the differential diagnosis of an infant or fetus with biventricular cardiac hypertrophy. It remains to be seen whether the masses in this case are an unusual manifestation of infantile Pompe, or represent the concurrence of tuberous sclerosis complex, or an otherwise isolated benign cardiac tumor.

Acknowledgments We would like to thank Dr. Paige Kaplan and Dr. Marc Yudkoff for their gracious assistance in editing this manuscript.

One-Sentence Take-Home Message

Cardiac masses may be an unusual presenting feature of the infantile form of glycogen storage disease type II, and may be identifiable on prenatal ultrasound evaluation.

References

Au KS, Williams AT, Roach ES et al (2007) Genotype/phenotype correlation in 325 individuals referred for a diagnosis of tuberous sclerosis complex in the United States. Genet Med 9:88–100

Barker PC, Pasquali SK, Darty S et al (2010) Use of cardiac magnetic resonance imaging to evaluate cardiac structure, function and fibrosis in children with infantile Pompe disease on enzyme replacement therapy. Mol Genet Metab 101:332–337

Dabora SL, Jozwiak S, Franz DN et al (2001) Mutational analysis in a cohort of 224 tuberous sclerosis patients indicates increased severity of TSC2, compared with TSC1, disease in multiple organs. Am J Hum Genet 68:64–80

De Cobelli F, Esposito A, Belloni E et al (2009) Delayed-enhanced cardiac MRI for differentiation of Fabry's disease from symmetric hypertrophic cardiomyopathy. AJR Am J Roentgenol 192: W97–W102

Ehlers KH, Hagstrom JW, Lukas DS, Redo SF, Engle MA (1962) Glycogen-storage disease of the myocardium with obstruction to left ventricular outflow. Circulation 25:96–109

Hohn AR, Lowe CU, Sokal JE, Lambert EC (1965) cardiac problems in the glycogenoses with specific reference to Pompe's disease. Pediatrics 35:313–321

Jones AC, Shyamsundar MM, Thomas MW et al (1999) Comprehensive mutation analysis of TSC1 and TSC2-and phenotypic correlations in 150 families with tuberous sclerosis. Am J Hum Genet 64:1305–1315

Kishnani PS, Hwu WL, Mandel H, Nicolino M, Yong F, Corzo D (2006) A retrospective, multinational, multicenter study on the natural history of infantile-onset Pompe disease. J Pediatr 148:671–676

Rees A, Elbl F, Minhas K, Solinger R (1976) Echocardiographic evidence of outflow tract obstruction in Pompe's disease (glycogen storage disease of the heart). Am J Cardiol 37:1103–1106

Sancak O, Nellist M, Goedbloed M et al (2005) Mutational analysis of the TSC1 and TSC2 genes in a diagnostic setting: genotype–phenotype correlations and comparison of diagnostic DNA techniques in tuberous sclerosis complex. Eur J Hum Genet 13:731–741

Shapir Y, Roguin N (1985) Echocardiographic findings in Pompe's disease with left ventricular obstruction. Clin Cardiol 8:181–185

Verhoef S, Bakker L, Tempelaars AM et al (1999) High rate of mosaicism in tuberous sclerosis complex. Am J Hum Genet 64:1632–1637

JIMD Reports
DOI 10.1007/8904_2011_87

The Use of Elevated Doses of Genistein-Rich Soy Extract in the Gene Expression-Targeted Isoflavone Therapy for Sanfilippo Disease Patients

Věra Malinová · Grzegorz Węgrzyn · Magdalena Narajczyk

Received: 11 May 2011 / Revised: 26 June 2011 / Accepted: 25 August 2011 / Published online: 11 December 2011
© SSIEM and Springer-Verlag Berlin Heidelberg 2011

Abstract Mucopolysaccharidoses (MPS) are severe, inherited metabolic disorders caused by storage of glycosaminoglycans (GAGs). Sanfilippo disease (mucopolysaccharidosis type III, MPS III) is described as severe neurological type of MPS, characterized by rapid deterioration of brain functions. No therapy for Sanfilippo disease is approved to date, however, a specific substrate reduction therapy (SRT), called gene expression-targeted isoflavone therapy (GET IT), has been used as an experimental therapy. In this report, we describe effects of treatment of six Sanfilippo disease patients with GET IT, in which the dose of genistein (5,7-dihydroxy-3-(4-hydroxyphenyl)-$4H$-1-benzopyran-4-one), an active compound of GET IT present in the soy isoflavone extract, has been increased to 10, and then to 15 mg/kg/day, contrary to the previously reported dose of 5 mg/kg/day. By measuring levels of urinary GAGs and assessing hair dysmorphology as biomarkers, and by considering clinical symptoms of patients, we obtained results suggesting that elevated doses of genistein may improve efficacy of GET IT for Sanfilippo disease.

Introduction

A large fraction of metabolic brain diseases consists of inherited disorders. Mucopolysaccharidoses (MPS) are a group of inherited metabolic disorders, caused by genetic defects resulting in accumulation of undegraded glycosaminoglycans (GAGs) in lysosomes of patients' cells (Beck 2007; Neufeld and Muenzer 2001). If heparan sulfate (HS) is one of the accumulated GAGs, severe symptoms occur in the central nervous system (CNS), including rapid deterioration of brain functions (Wegrzyn et al. 2010a). Sanfilippo disease (mucopolysaccharidosis type III or MPS III) is characterized by the sole accumulation of HS, and the brain dysfunction-related symptoms are especially severe in this MPS type (reviewed by Valstar et al. 2008).

Until now, no therapy has been approved for treatment of Sanfilippo disease patients. However, a specific kind of substrate reduction therapy (SRT), called gene expression-targeted isoflavone therapy (GET IT), has been proposed to be used in treatment of MPS patients, especially those suffering from brain dysfunctions. Genistein (5,7-dihydroxy-3-(4-hydroxyphenyl)-$4H$-1-benzopyran-4-one) was demonstrated to be an inhibitor of GAG synthesis in fibroblasts of patients suffering from various MPS types, including MPS III (Piotrowska et al. 2006). Subsequent studies indicated that genistein may be effective in treatment of animal models of MPS II (Friso et al. 2010) and MPS IIIB (one of four subtypes of MPS III) (Malinowska et al. 2009). Importantly, a complete correction of behavior of MPS

Communicated by: Verena Peters.

Competing interests: None declared.

V. Malinová
Department of Pediatrics and Adolescent Medicine, Charles University, Ke Karlovu 2
120 00, Praha 2, Czech Republic

G. Węgrzyn
Department of Molecular Biology, University of Gdańsk, Kładki 24, 80-822, Gdańsk, Poland

M. Narajczyk (✉)
Laboratory of Electron Microscopy, University of Gdańsk, Kładki 24, 80-822, Gdańsk, Poland
e-mail: magnaraj@biotech.ug.gda.pl

IIIB mice was observed in long-term treatment of animals with a relatively high dose of genistein (Malinowska et al. 2010).

The encouraging results of experiments performed in vitro and on animal models led to the proposal that GET IT may be a hopeful option for treatment of neuronopathic forms of MPS (for see Piotrowska et al. (2006), for discussions see Wegrzyn et al. (2010b) and de Ruijter et al. (2011)). Subsequent, more detailed in vitro studies confirmed that genistein and other flavonoids are potent inhibitors of GAG synthesis, and their low cytotoxicity and a potential to cross the blood–brain barrier confirmed that GET IT should be tested as a putative therapy (Arfi et al. 2010; Piotrowska et al. 2010; Kloska et al. 2011). In fact, a pilot (open-label) clinical study with the use of a genistein-rich isoflavone extract for treatment of Sanfilippo patients has been performed and the results were encouraging, including statistically significant reduction of urinary GAG levels (though the reduction occurred in seven out of ten investigated patients), improvement in hair morphology (in eight patients), and an increase in the score achieved in a psychological test (in eight patients) (Piotrowska et al. 2008). In that study, the genistein dose was 5 mg/kg/day and the duration of the treatment was 1 year. Then, a 2-year follow-up study was performed, indicating that at this particular dose of genistein, after the initial improvement, the patients' state stabilized (in some patients) or slowly deteriorated (in other patients) (Piotrowska et al. 2011). Another recent study showed that GET IT may be useful in treatment of patients suffering from MPS II (Hunter disease), another type of MPS in which a large fraction of patients suffers from deterioration of brain functions (Marucha et al. 2011).

In all studies on experimental GET IT for MPS patients published to date, a genistein-rich isoflavone extract was used with the genistein dose of 5 mg/kg/day (Piotrowska et al. 2008, 2011; Delgadillo et al. 2011; Marucha et al. 2011). However, authors of the most recent studies suggested that an increase in the genistein dose might result in higher efficacy of GET IT (Delgadillo et al. 2011; Piotrowska et al. 2011). Therefore, we aimed to study effects of this experimental therapy when the genistein dose is initially doubled (10 mg/kg/day) and then, after several months, increased up to 15 mg/kg/day. Assessment of previously established biomarkers, urinary GAG levels, and hair morphology, as well as results of clinical observations, were employed to evaluate effects of the treatment.

Materials and Methods

Patients

The patients were diagnosed for Sanfilippo disease (either MPS IIIA, McKusick's OMIM no. 252900, or MPS IIIB, McKusick's OMIM no. 252920) by estimation of urinary GAG levels and measurement of activities of particular lysosomal hydrolases in leukocytes, according to the optimized methods described previously (Piotrowska et al. 2008). Deficiency in activity of heparan N-sulfatase (control value: 4.1 ± 1.4 nmol/mg of protein/18 h) or α-N-acetyl glucosaminidase (control value: 90 ± 34 nmol/mg of protein/42 h) was considered as a diagnosis for MPS IIIA or MPS IIIB, respectively. All patients were of Caucasian origin and came from the area of Czech Republic. Among them, there were three males and three females. Information about all patients enrolled into this study is summarized in Table 1.

Treatment and Assessment of Its Effects

The MPS III patients, characterized in Table 1, were treated for the period of over 1 year, in the range between 15 and 24 months (the differences arose from various times of the enrollment of patients into the study, which resulted from the rarity of the disease and the attempt to include each patient to the study as soon as possible due to ethical issues, namely a lack of an alternative treatment of this severe disease). For the treatment, a genistein-rich soy isoflavone extract (called SE-2000 or Soyfem), provided by the manufacturer (Biofarm, Poznań, Poland) in the form of

Table 1 Characteristics of patients

Patient no.	MPS type	Gender	Age at diagnosis[a]	Age at the therapy onset[a]
1	MPS IIIA	Female	0 y 10 m	7 y 0 m
2	MPS IIIA	Male	7 y 10 m	11 y 1 m
3	MPS IIIB	Female	5 y 0 m	6 y 6 m
4	MPS IIIA	Female	3 y 6 m	3 y 6 m
5	MPS IIIA	Male	1 y 8 m	2 y 0 m
6	MPS IIIA	Male	3 y 0 m	4 y 7 m

[a] Age is provided in years (y) and months (m)

tablets (the product named Soyfem), was used. The extract consists of genistin and genistein (26.90%), daidzin and daidzein (13.37%), glycitin and glycitein (1.98%), and soy proteins, carbohydrates, and lipids (remaining amount). This extract was administered orally (in the form of whole tablets or tablets crushed into powder) at the initial dose corresponding to 10 mg of genistin and genistein (genistin is a glycan that can be converted to genistein by either acid environment or intestinal bacteriological flora) per 1 kg of body weight daily, which was then (after several months of the treatment) increased to 15 mg/kg/day. The extract was administered four times a day, with equal amounts each time. To assess the effects of the treatment, two biomarkers were measured: (1) urinary GAG levels and (2) hair dysmorphology (the investigator was blinded in the process of sample analysis), which were measured according to the previously published methods (Malinowska et al. 2008; Piotrowska et al. 2008). Clinical symptoms of patients were assessed by their basic investigation and observation during

visits in the clinic, as well as by analysis of interviews with patients' parents. The monitoring of adverse effects was based on reports of parents, who were appropriately instructed to signal any such effects immediately (by either phone or electronic mail, with confirmation of receipt of the information), and who provided an assessment of such effects in a written form every 3 months (even if no adverse effects were observed). This experimental treatment has been approved by the State Institute for Drug Control (Prague, Czech Republic). Parents of the children involved in this study have signed informed consent form.

Results and Discussion

The results of the treatment of six patients suffering from Sanfilippo disease with GET IT, assessed by monitoring two biomarkers, urinary GAG levels, and hair dysmorphology, are presented in Fig. 1. Initially, patients were treated

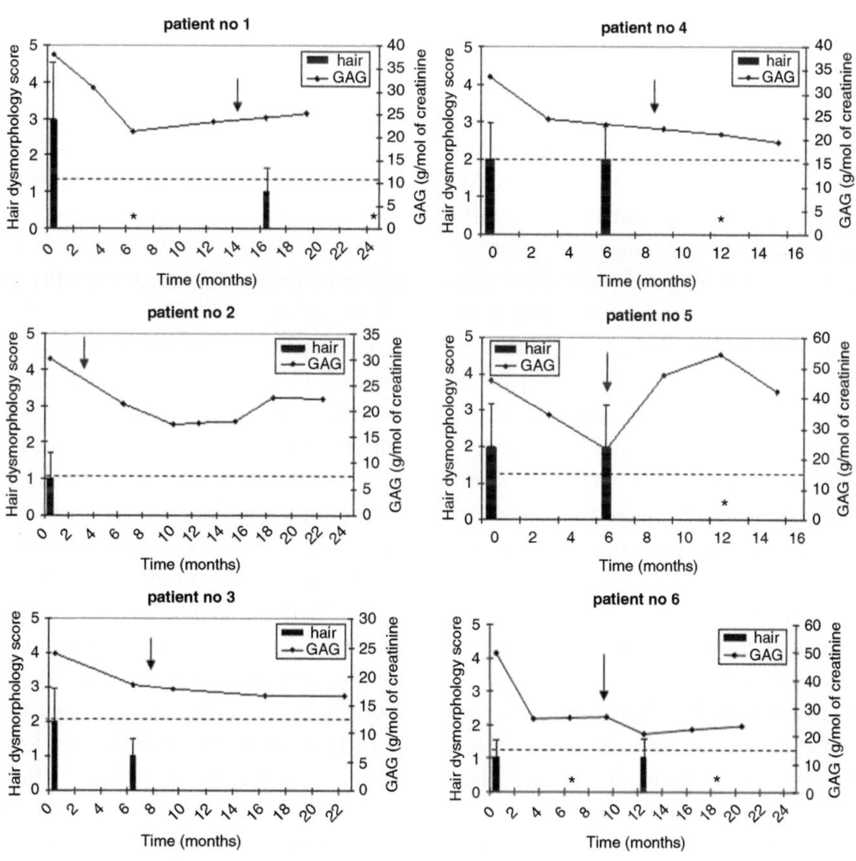

Fig. 1 Effects of GET IT in six Sanfilippo disease patients as assessed by measuring urinary GAG levels and hair dysmorphology as biomarkers. Numbers of patients correspond to those provided in Table 1. The patients were treated with genistein-rich soy isoflavone extract from time = 0 with the dose corresponding to the amount of genistein equal to 10 mg/kg/day, and the dose was elevated to 15 mg/kg/day at the time indicated by *arrow*. Urinary GAG levels (in g/mol of creatinine) are shown as diagrams, with the age-specific upper norm value represented by a *dashed horizontal line*. Level of hair dysmorphology is presented in the form of *histograms*, with average values from at least ten independent estimations and *error bars* representing standard deviation (SD); *asterisks* represent investigated samples for which the value was 0 with 0 SD; the scale is from 0 (normal hair morphology) to 5 (the most abnormal hair morphology), according to Malinowska et al. (2008)

Table 2 Clinical changes in patients during GET IT as assessed by: (a) basic investigation and observation of patients during their visits in the clinic, and (b) interviews with patients' parents

Patient no.	Clinical observations	
	Major symptoms at baseline	Changes from baseline to the endpoint
1	Facial dysmorphology; hepatosplenomegaly; frequent infections; frequent falls; epileptic attacks; severe mental retardation with regression of development; autistic features of behavior; sleep problems; hyperactivity	Ability to walk without support; no epileptic attacks since the therapy onset; improved reactions to surrounding stimulation; improved sleep; decreased hyperactivity
2	Facial dysmorphology; hepatosplenomegaly; mental retardation with regression of development	Mild improvement in communication skills; improved response to surrounding signs (these changes were observed only after the dose increase)
3	Facial dysmorphology; frequent falls; epileptic attacks; mental retardation with regression of development; attacks of aggression and self-injuries	Inhibition of developmental regression (observed only after the dose increase)
4	Hepatomegaly; delay in motor skills; slow regression in speech and social adaptation; sleep problems; increased agitation	Inhibition of developmental regression; improvement in speech (more words used and better pronunciation); decreased agitation (these changes were observed only after the dose increase)
5	Hepatosplenomegaly; frequent falls; developmental regression; impaired speech; sleep problems; hyperactivity	Inhibition of developmental regression; improved sleep; decreased hyperactivity (these changes were observed only after the dose increase)
6	Hepatosplenomegaly; developmental regression; sleep problems; hyperactivity	Improved sleep; decreased hyperactivity

with the genistein dose of 10 mg/kg/day (from time 0), and this treatment resulted in a decrease in the urinary GAG level in all patients (this decrease was statistically significant, $p < 0.05$, when the analysis of results was performed considering all patients as an investigated population; data not shown), though in none of them the level dropped to the norm value (Fig. 1). However, at this dose, an improvement in hair dysmorphology was observed only in three patients (patients no. 1, 3, and 6). After various times of such a treatment, the dose of genistein has been increased to 15 mg/kg/day (the time of this increase is marked by arrow in Fig. 1). The exact time of the dose change was chosen on the basis of the clinical status of particular patient, and represented either at the end of the period of the plateau of clinical symptoms observed between two subsequent visits in the clinic after initial improvement or after a lack of improvement seen between the first and the second visit. In most patients, this change had either minor or not significant effects on the further decrease of urinary GAG levels, however, after the dose increase, the level of hair dysmorphology decreased significantly and reached the norm value in all patients (Fig. 1). Unexpected results of urinary GAG levels and hair morphology were obtained in patient no. 5, and patients no. 1 and 6, respectively, after the dose increase. Namely, the GAG level increased and hair morphology deteriorated transiently, despite significant initial improvements in the values of these biomarkers. The reason(s) for these phenomena remain(s) unknown, but we cannot exclude an unspecific reaction to any unidentified conditions, other than the primary disease.

Clinical improvement was observed in all patients, but to various extent (Table 2). In patients no. 2, 3, 4, and 5, the clinical effects were noted only after the dose increase up to 15 mg/kg/day (Table 2).

Generally, the results presented in Fig. 1 and Table 2 indicate that GET IT with the use of elevated doses of genistein (10 and then 15 mg/kg/day) relative to that administered in previous studies (5 mg/kg/day, Piotrowska et al. 2008, 2011; Delgadillo et al. 2011; Marucha et al. 2011) improved both the values of biomarkers and clinical parameters used for assessment of the efficacy of this treatment. Contrary to the use of the lower dose of genistein (5 mg/kg/day), in this study, we observed a decrease in urinary GAG levels in all (not only in some) patients treated with the dose of 10 mg/kg/day, and a further increase in the dose, up to 15 mg/kg/day, correlated with normalization of hair morphology in all patients. This may suggest an improved efficacy after the dose increase, however, we cannot exclude that the hair morphology normalization was caused by a long exposure to genistein rather than to the use of the increased dose. Although some clinical improvement was observed in all patients, in four out of six patients, the positive changes could be noted only during the treatment with genistein at 15 mg/kg/day. Therefore, we suggest that the proposals to increase the dose of genistein in GET IT for Sanfilippo disease, published recently (Delgadillo et al. 2011; Piotrowska et al. 2011), were substantiated, though further studies on efficacy of GET IT are undoubtedly required.

Since the best results in correction of the behavior of mice suffering from MPS IIIB have been obtained with the use of the dose of genistein as high as 160 mg/kg/day (Malinowska et al. 2010), it remains to be determined whether further increase in the genistein dose may further improve the efficacy of GET IT in humans. Safety of the treatment should be an important issue in such putative studies. No adverse effects were reported by parents of the patients investigated in this study, which confirmed previous observations (Piotrowska et al. 2008, 2011; Delgadillo et al. 2011; Marucha et al. 2011) that this treatment is safe. However, it is clear that to achieve the genistein dose similar to that used in experiments with mice, it will not be possible to provide a genistein-rich soy isoflavone extract. Rather, pure (either purified or synthetic) genistein should be used to eliminate potential adverse effects caused by the compounds present in the extracts along with genistein.

In conclusion, it appears that the use of increased doses of genistein may have positive effects on efficacy of GET IT in treatment of patients suffering from Sanfilippo disease. However, further studies, including double-blinded placebo-controlled clinical trials with various genistein doses, are required to unambiguously assess efficacy and to optimize the treatment procedures.

Acknowledgments This research was supported by Ministry of Sciences and Higher Education of Poland (project grant no. N N301 668540 to GW), and was operated within the Foundation for Polish Science Team Programme co-financed by the EU European Regional Development Fund (grant no. TEAM/2008-2/7 to GW). A support from Polish MPS Society is greatly acknowledged.

Conflict of Interest

The authors declare no conflict of interest.

References

Arfi A, Richard M, Gandolphe C, Scherman D (2010) Storage correction in cells of patients suffering from mucopolysaccharidoses types IIIA and VII after treatment with genistein and other isoflavones. J Inherit Metab Dis 33:61–67

Beck M (2007) Mucopolysaccharidoses: clinical features and management. In: vom Dahl S, Wendel U, Strohmeyer G (eds) Genetic metabolic disorders: management, costs and sociomedical aspects.. Deutscher Arzte-Verlag, Cologne, pp 13–18

de Ruijter J, Valstar MJ, Wijburg FA (2011) Mucopolysaccharidosis type III (Sanfilippo syndrome): emerging treatment strategies. Curr Pharm Biotechnol 12(6):923–930

Delgadillo V, del Mar O'Callaghan M, Artuch R, Montero R, Pineda M (2011) Genistein supplementation in patients affected by Sanfilippo disease. J Inherit Metab Dis 34(5):1039–1044, doi:10.1007/s10545-011-9342-4

Friso A, Tomanin R, Salvalaio M, Scarpa M (2010) Genistein reduces glycosaminoglycan levels in a mouse model of mucopolysaccharidosis type II. Br J Pharmacol 159:1082–1091

Kloska A, Jakóbkiewicz-Banecka J, Narajczyk M, Banecka-Majkutewicz Z, Węgrzyn G (2011) Effects of flavonoids on glycosaminoglycan synthesis: implications for substrate reduction therapy in Sanfilippo disease and other mucopolysaccharidoses. Metab Brain Dis 26:1–8

Malinowska M, Jakóbkiewicz-Banecka J, Kloska A, Tylki-Szymańska A, Czartoryska B, Piotrowska E, Wegrzyn A, Wegrzyn G (2008) Abnormalities in the hair morphology of patients with some but not all types of mucopolysaccharidoses. Eur J Pediatr 167:203–209

Malinowska M, Wilkinson FL, Bennett W et al (2009) Genistein reduces lysosomal storage in peripheral tissues of mucopolysaccharide IIIB mice. Mol Genet Metab 98:235–242

Malinowska M, Wilkinson FL, Langford-Smith KJ et al (2010) Genistein improves neuropathology and corrects behaviour in a mouse model of neurodegenerative metabolic disease. PLoS One 5:e14192

Marucha J, Tylki-Szymańska A, Jakóbkiewicz-Banecka J, Piotrowska E, Kloska A, Czartoryska B, Węgrzyn G (2011) Improvement in the range of joint motion in seven patients with mucopolysaccharidosis type II during experimental gene expression-targeted isoflavone therapy (GET IT). Am J Med Genet A 155A (9):2257–2262

Neufeld EF, Muenzer J (2001) The mucopolysaccharidoses. In: Scriver CR, Beaudet AL, Sly WS, Valle D (eds) The metabolic and molecular bases of inherited disease. McGraw-Hill Co., New York, pp 3421–3452

Piotrowska E, Jakobkiewicz-Banecka J, Baranska S et al (2006) Genistein-mediated inhibition of glycosaminoglycan synthesis as a basis for gene expression-targeted isoflavone therapy for mucopolysaccharidoses. Eur J Hum Genet 14:846–852

Piotrowska E, Jakobkiewicz-Banecka J, Tylki-Szymanska A et al (2008) Genistin-rich soy isoflavone extract in substrate reduction therapy for Sanfilippo syndrome: An open-label, pilot study in 10 pediatric patients. Curr Ther Res Clin Exp 63:166–179

Piotrowska E, Jakóbkiewicz-Banecka J, Wegrzyn G (2010) Different amounts of isoflavones in various commercially available soy extracts in the light of gene expression-targeted isoflavone therapy. Phytother Res 24(Suppl 1):S109–S113

Piotrowska E, Jakobkiewicz-Banecka J, Maryniak A, Tylki-Szymanska A, Puk E, Liberek A, Wegrzyn A, Czartoryska B, Slominska-Wojewodzka M, Wegrzyn G (2011) Two-year follow-up of Sanfilippo Disease patients treated with a genistein-rich isoflavone extract: Assessment of effects on cognitive functions and general status of patients. Med Sci Monit 17:CR196–CR202

Valstar MJ, Ruijter GJ, van Diggelen OP, Poorthuis BJ, Wijburg FA (2008) Sanfilippo syndrome: a mini-review. J Inherit Metab Dis 31:240–252

Wegrzyn G, Jakóbkiewicz-Banecka J, Narajczyk M, Wiśniewski A, Piotrowska E, Gabig-Cimińska M, Kloska A, Słomińska-Wojewódzka M, Korzon-Burakowska A, Węgrzyn A (2010a) Why are behaviors of children suffering from various neuronopathic types of mucopolysaccharidoses different. Med Hypoth 75:605–609

Wegrzyn G, Jakóbkiewicz-Banecka J, Gabig-Cimińska M, Piotrowska E, Narajczyk M, Kloska A, Malinowska M, Dziedzic D, Gołebiewska I, Moskot M, Wegrzyn A (2010b) Genistein: a natural isoflavone with a potential for treatment of genetic diseases. Biochem Soc Trans 38:695–701

JIMD Reports
DOI 10.1007/8904_2011_88

RESEARCH REPORT

Pregnancy During Nitisinone Treatment for Tyrosinaemia Type I: First Human Experience

A. Vanclooster · R. Devlieger · W. Meersseman ·
A. Spraul · K. Vande Kerckhove · P. Vermeersch ·
A. Meulemans · K. Allegaert · D. Cassiman

Received: 27 June 2011 / Revised: 19 July 2011 / Accepted: 26 August 2011 / Published online: 16 December 2011
© SSIEM and Springer-Verlag Berlin Heidelberg 2011

Abstract A 19 year old woman with tyrosinaemia type 1 gave birth to a healthy girl after 41 weeks of gestation. Nitisinone was continued throughout the pregnancy (maternal levels 68–96 μmol/l, target level 30–60 μmol/l).

Communicated by: Claude Bachmann.

Competing interests: None declared.

A. Vanclooster (✉) · W. Meersseman · K.V. Kerckhove ·
P. Vermeersch · A. Meulemans · D. Cassiman
Department of Pediatrics and Metabolic Center, University
Hospital Gasthuisberg, University of Leuven, Herestraat 49,
3000, Leuven, Belgium
e-mail: annick.vanclooster@uzleuven.be

W. Meersseman
e-mail: wouter.meersseman@uzleuven.be

K.V. Kerckhove
e-mail: kristel.vandekerckhove@uzleuven.be

P. Vermeersch
e-mail: pieter.vermeersch@uzleuven.be

A. Meulemans
e-mail: ann.meulemans@uzleuven.be

D. Cassiman
e-mail: david.cassiman@med.kuleuven.be

R. Devlieger
Department of Obstetrics and Gynecology, University Hospitals
K.U.Leuven, Leuven, Belgium
e-mail: roland.devlieger@uzleuven.be

A. Spraul
Biochemistry Unit, CHU Bicêtre, Assistance Publique-Hôpitaux
de Paris, Paris, France
e-mail: anne.spraul@bct.ap-hop-paris.fr

K. Allegaert
Department of neonatology, University Hospitals K.U.Leuven,
Leuven, Belgium
e-mail: karel.allegaert@uzleuven.be

Tyrosine levels during pregnancy were between 500 and 693 μmol/l (normal values 20–120 μmol/l) and phenylalanine levels between 8 and 39 μmol/l (normal values 30–100 μmol/l). Nitisinone was measurable in neonatal blood immediately after birth, at a level comparable to the simultaneous level in the mother. Nitisinone half-life in the neonate was estimated to be 90 h. Tyrosine levels in the neonate decreased from 1,157 μmol/l at birth (cord blood) to normal levels within 4 weeks. Phenylalanine levels in the neonate were normal from birth on. The child had a normal psychomotor development as assessed throughout the first year of life.

This is the first report worldwide of a pregnancy during treatment with nitisinone.

In this case, no adverse effects of nitisinone, maternal high tyrosine or low phenylalanine were detected in the child, so far. Long-term results in a larger cohort of pregnancies and births are needed to determine whether nitisinone can be administered safely during pregnancy.

Abbreviations

HT-1 Hepatorenal tyrosinaemia
Phe Phenylalanine
Tyr Tyrosine

Introduction

Tyrosinaemia type 1 (hepatorenal tyrosinaemia, HT-1, OMIM: 276700) is a rare, metabolic disorder affecting about one child in 100,000. It is a hereditary autosomal recessive disease caused by a deficiency in the enzyme fumarylacetoacetase (FAH). Patients often present with liver failure and gastrointestinal bleeding (Scott 2006).

Renal signs of HT-1 in infants include tubulopathy and Fanconi syndrome. Due to renal phosphate wasting, rickets can develop (Scott 2006; Kvittingen and Holme 2000).

Nitisinone (NTBC; 2-(2-nitro-4-trifluoro-methylbenzoyl)-1,3-cyclohexanedione) reverses the lethal nature of the disease by preventing the accumulation of the toxic metabolites, maleylacetoacetate, and fumarylacetoacetate, hence allowing HT-1 patients to lead a normal life. Nitisinone prevents the accumulation of fumarylacetoacetate and, therefore, its conversion to succinylacetone (Grompe 2001).

Animal toxicity studies with nitisinone have shown teratogenicity. A dose 2.5-fold higher than the maximum recommended human dose (2 mg/kg/day) caused abdominal wall defects (omphalocoele and gastroschisis) in rabbits. A second study, in mice, showed a statistically significant reduced survival and growth of the pups during the weaning period at dose levels of 125- and 25-fold, respectively, the maximum recommended human dose. Rats showed a reduced mean pup weight and corneal lesions due to exposure to nitisinone through breast milk (Summary of product characteristics Orfadin; http://www.ema.europa.eu).

Three pregnancies in two mothers with tyrosinaemia type 2 have been reported so far. Tyrosinaemia type II patients show high tyrosinaemia, as well, but do not show the succinylacetone production of HT-1 patients. The first patient presented at 34 weeks of gestation with a plasma tyrosine (Tyr) level of 1,302 μmol/l and a phenylalanine (Phe) level of 37 μmol/l. The intake of protein ranged from 60 to 90 g per day. The first child was born at term and had a birth weight of 1.9 kg. He was evaluated at 1 year and 4 months and showed microcephaly and maxillary hypoplasia. Development testing showed a developmental quotient of 72 (Brunet–Lezine test). Her second child was evaluated at 12 months and also had microcephaly, was unable to walk, and had delayed speech development. Both children had normal Tyr levels at the time of evaluation (Cerone et al. 2002).

In the second report, pregnancy in a 25 year old woman with tyrosinaemia type II is described. A protein restricted diet supplemented with a Tyr/Phe free amino acid mixture was followed from week 5 of pregnancy. A restriction of natural protein intake to 0.16 g/kg/day in early pregnancy and 0.38 g/kg/day in the last trimester reduced Tyr levels to between 100 and 200 μmol/l. Plasma Phe levels stayed in the range of 20–40 μmol/l. There was a normal weight gain and fetal growth and the mother remained asymptomatic throughout pregnancy. The neonate was born at term with length, weight, and head circumference between the twenty-fifth and fiftieth percentiles (Francis et al. 1992).

Materials and Methods

Amino acids and succinylacetone on blood spots of the newborn were determined by tandem-MS. Amino acids in the cord blood of the newborn and in the mother were determined by ion-exchange chromatography.

Nitisinone levels in plasma were measured with an UPLC-MSMS-based technique. A nitisinone concentration between 30 and 60 μmol/l is considered to be therapeutic and minimizes the production of succinylacetone in the urine (Davit-Spraul et al. 2011).

The blood samples of the newborn taken on day 0 were from the umbilical cord. All other blood samples are capillary (blood spot).

Case Report

Personal and Family History

A girl born in 1990 from non-consanguineous parents was diagnosed with HT-1 at the age of 3 years when she presented with rickets, hepatosplenomegaly, phosphaturia, and tubulopathy (hyperaminoaciduria, hypercalciuria, and glucosuria). Liver biopsy showed micro–macronodular cirrhosis and kidney biopsy revealed tubular damage. An ultrasound showed enlarged kidneys. She was started on nitisinone in January 1994 and in February 1995 blood α-fetoprotein and succinylacetone in urine and blood had normalized.

Over the years, all biochemical parameters remained stable, there was no proteinuria, no urinary succinylacetone was detected since the start of nitisinone. Apart from her intake of nitisinone which was very accurate, the protein restricted diet, supplemented with Phe/Tyr free amino acid mixture, was followed correctly. On abdominal MRI, the liver margins were irregular, but there were no indications of cirrhosis or portal hypertension, there were no focal lesions. The kidneys show small cysts.

In October 2009, at age 19, the patient was transferred to the adult metabolic clinic. She had no focal liver lesions, no cognitive dysfunction and a normal bone densitometry. Dietary adherence was very good (nitisinone 0.5 mg/kg/day). She was taking an oral contraceptive (Mercilon®, Organon Belgium NV; Belgium) since March 2009.

Levels of Tyr and Phe during the 4 years before pregnancy are shown in Fig. 1. As seen here, her Phe levels remained stable and within normal range (30–100 μmol/l). Tyr levels were elevated due to the underlying disorder and nitisinone intake; normal range: 20–120 μmol/l.

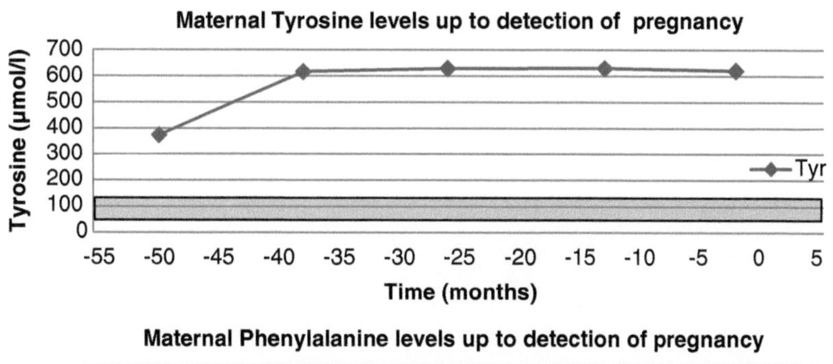

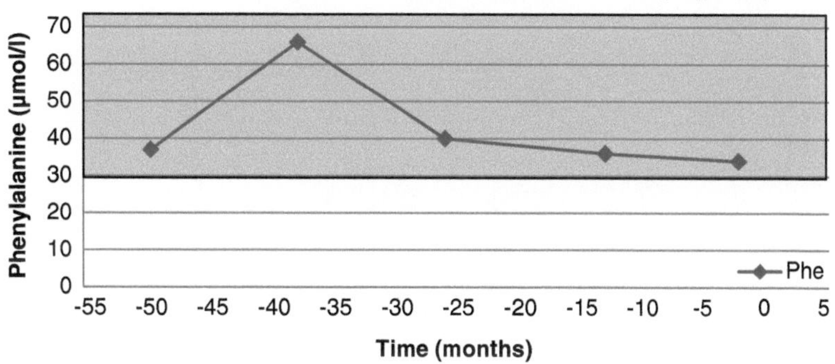

Fig. 1 Tyrosine and phenylalanine levels in the mother prior to detection of pregnancy; last value = at time of detection of pregnancy, at 6 months of pregnancy; date of delivery = time 0. Normal values are depicted as the *gray* shaded area

Pregnancy Course

Pregnancy was detected in January 2010, at 27 weeks of amenorrhea in an otherwise healthy HT-1 patient. The oral contraceptive pill was discontinued and detailed fetal imaging studies were performed. Ultrasound and MRI of the fetus were reassuring. Growth was within normal range for gestational age (Fig. 2). Discontinuation of nitisinone treatment was discussed with all involved parties. Nitisinone treatment 0.5 mg/kg/day (3 × 10 mg per day) was ultimately continued because pregnancy had developed without apparent problems so far and, therefore, the risk for hepatic or renal failure due to discontinuation of therapy was estimated to be more important than the risks related to nitisinone exposure to the fetus during the remaining of pregnancy. Figure 3 shows nitisinone levels during pregnancy. No values immediately prior to pregnancy are available.

Tyr and Phe levels were monitored regularly from the moment the pregnancy was detected. Tyr and Phe levels from 3 months prior to delivery until 9 months after delivery are presented in Fig. 3. At 33 weeks of pregnancy, there was a decrease of Phe (and Tyr) in plasma, with Phe decreasing below normal values. Dietary changes were considered necessary. The dose of amino acid mixture was elevated to 4 × 19 g Tyr2 (Phe/Tyr free amino acid mixture from Nutricia NV; Belgium) per day. This intake was spread over the day, three times with a meal and one intake before bedtime. Vitamin and mineral supplement (Omnibionta Pronatal, Merck Consumer Healthcare nv/sa; Belgium) was also introduced 2 months prior to delivery. Four protein free sandwiches per day were changed into two protein free sandwiches and two normal bread sandwiches, thus increasing the intake of natural protein. No more dietary changes were made during the last month of pregnancy.

Dietary details throughout the pregnancy are listed in Table 1.

Neonatal Course

At 41 weeks of gestation, labor was induced and a healthy girl was born after an uneventful delivery. Apgar score for breathing, pulse, and muscle tonus were scored and resulted in 0-0-1 at birth, 9-9-9 after 1 min and 10-10-10 after 5 min. The birth weight was 2,950 g, length was 49 cm.

Tyr, Phe, and nitisinone levels are shown in Fig. 4. Tyr in the newborn normalized slowly. After 28 days, an almost normal Tyr level was reached (50–150 µmol/l). The highest Phe level was measured on the day of birth but was within normal range (40–140 µmol/l) and fell rapidly the following days.

Nitisinone in the umbilical cord blood was 69 µmol/l, while in the mother it was 94 µmol/l at delivery (last intake

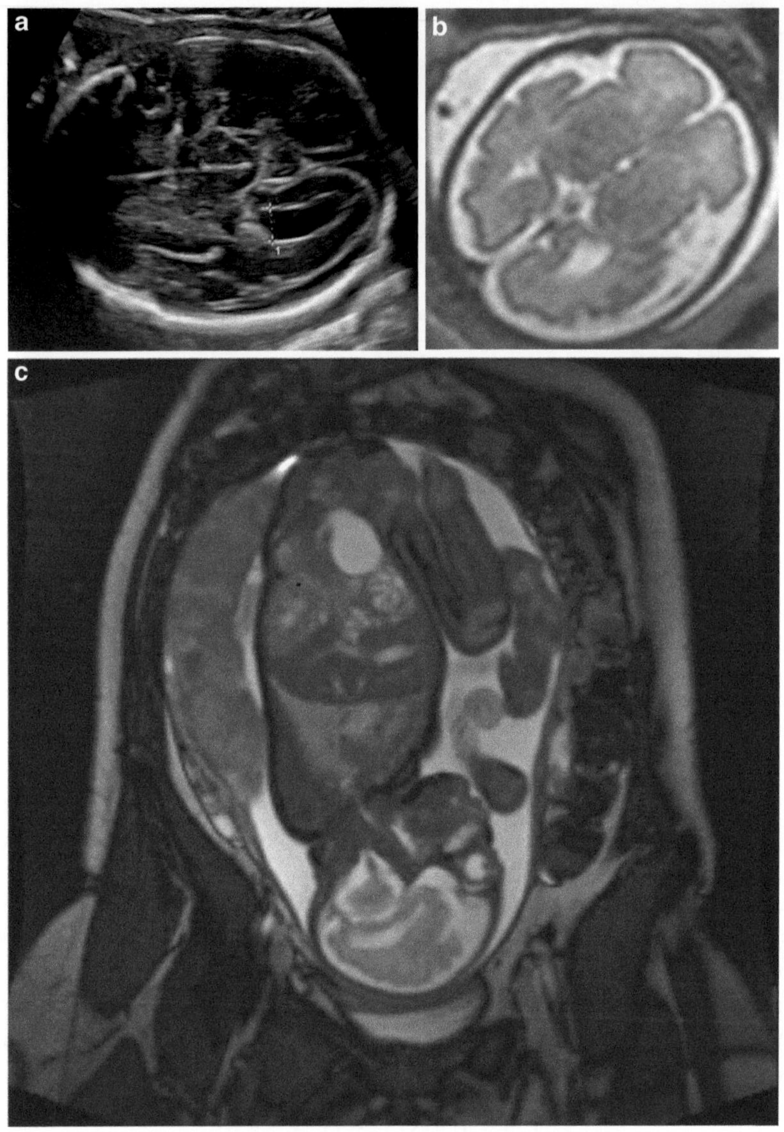

Fig. 2 (**a**) Prenatal ultrasound at 28 weeks of gestation. (**b**) Fetal MRI of the fetal brain at 34 weeks of gestation. Both imaging modalities show normal cerebral anatomy, gyration, and migration for the gestational age. (**c**) MRI scan of the fetus at 34 weeks of pregnancy shows normal development

of 10 mg of nitisinone 4 h prior to delivery). Five days after birth, the nitisinone level in the newborn was already under the target values and on day 28, no nitisinone could be detected (Fig. 4). There was no succinylacetone detected in the blood spots of the newborn at any time point. The calculated half-life (elimination half-life) is 90 h, this is based on two observations, thus it is an estimation. We presumed a single compartment model, with linear elimination ($y = 69 + 0.3833x$).

The mother was not keen on breastfeeding the child and this choice was supported. Data showed a reduced mean weight and corneal lesions in rat pups due to nitisinone exposure via milk (Summary of product characteristics Orfadin; http://www.ema.europa.eu).

Weight and length of the newborn were within normal standards for the Flemish population (SPE standards; Devlieger et al. 1996). Postnatal growth in the child follows the p50–75 (Percentile 50–75) and her head circumference follows the p10–25 up till 12 months of age (i.e., to date). Her mother, during childhood, followed p50–75 for weight and p25–50 for length and head circumference. Data of the father are not available. The child's hearing exam was normal at the age of 2 weeks.

At approximately 10 months, the child underwent an extensive neurological examination which showed no abnormalities. The mental development score was 97 and motor development score 86 (Bayley Scales of Infant Development) which is within normal range. In general,

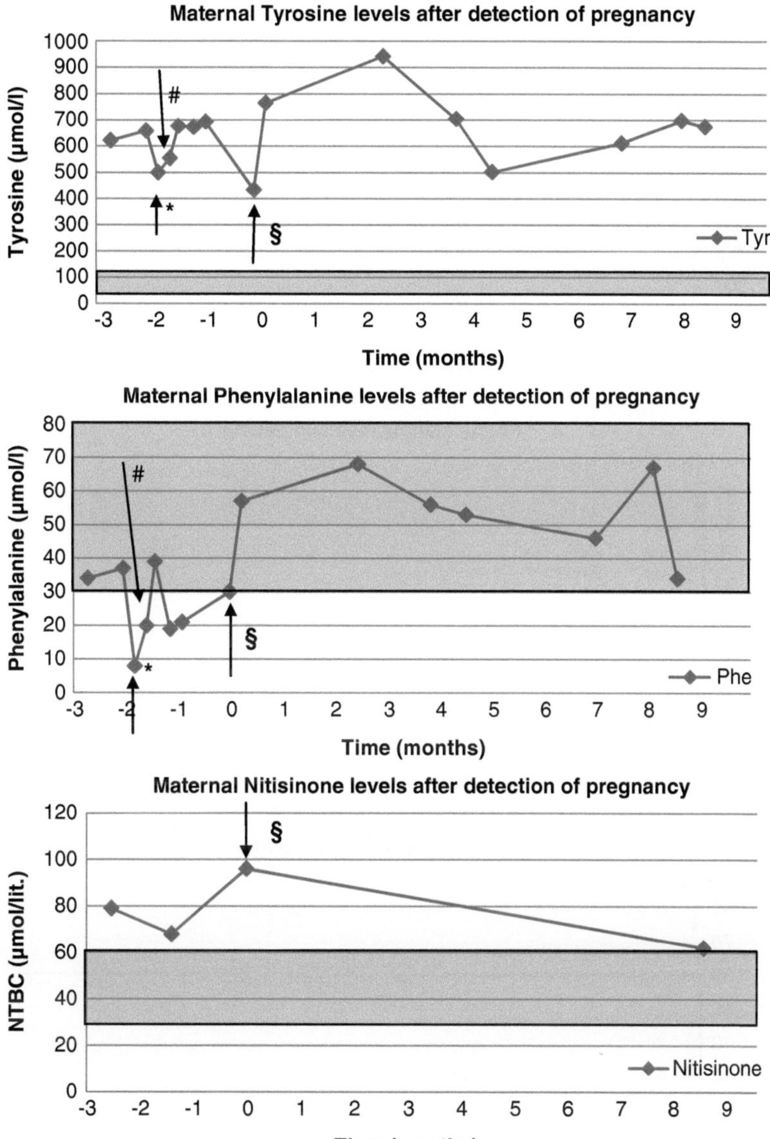

Fig. 3 Tyrosine, phenylalanine, and nitisinone levels in the mother; * switch from 3 × 15 g Tyr 2 to 4 × 19 g Tyr 2; # starting two normal sandwiches; § date of delivery (time = 0). Normal values are depicted as the *gray* shaded areas

Table 1 Diet during pregnancy

Time of pregnancy	Energy intake (kcal/day)	Protein intake (g protein/ kg bodyweight/day)	Natural protein intake (g protein/kg/day)	Protein intake from amino acid supplement (tyr 2) (g protein/kg/day)	Energy
Prior to pregnancy until 32 weeks of pregnancy	2,319	55.6 g P/day = 0.8 g P/ day/kg	0.4	0.4	9.7% P32.4% F57.9% CH
33 weeks	2,524	71.1 g P/day = 0.9 g P/ kg/day	0.3	0.6	11.5% P29.4% F59.1% CH
34 weeks	2,488	76 g P/day = 1 g P/kg/ day	0.4	0.6	12,3% P30.1% F57,6% CH
41 weeks	2,971	75 g P/day = 1 g P/kg/ day	0.4	0.6	10.2% P28.1% F61.7% CH

P = protein, F = fat, and CH = carbohydrate

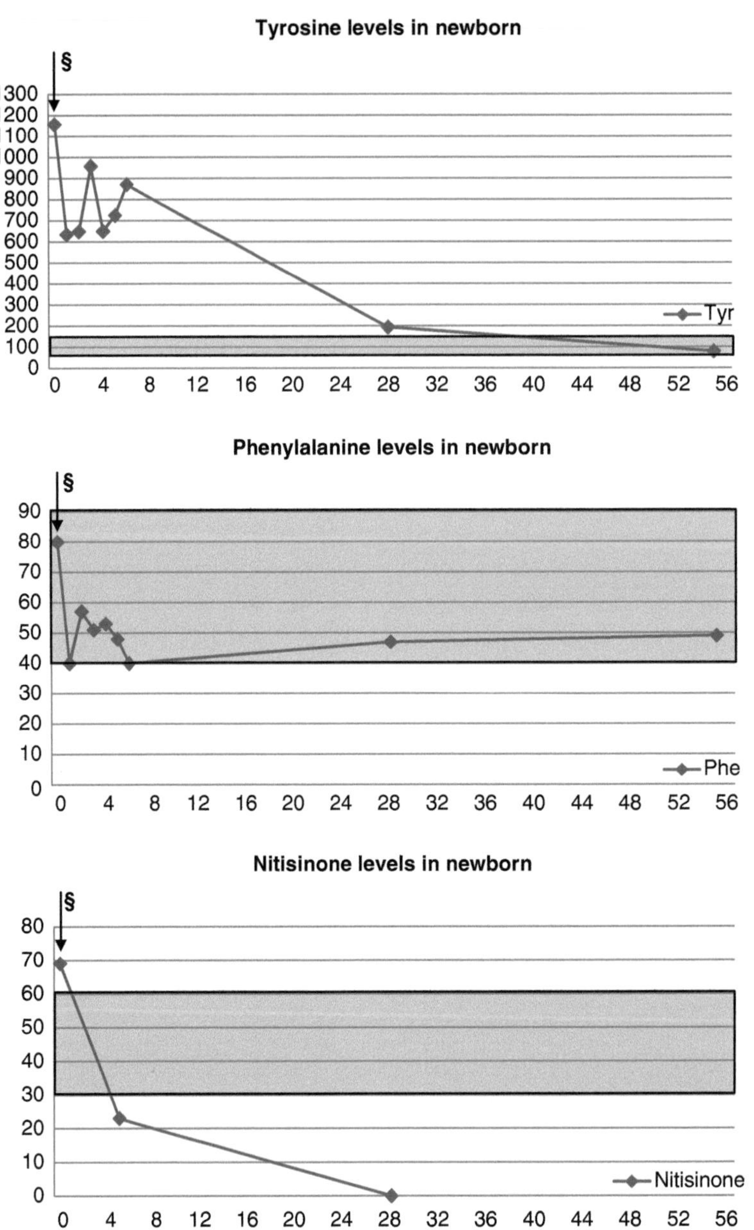

Fig. 4 Tyrosine (1157 μmol/l at birth, cord blood), phenylalanine (80 μmol/l at birth, cord blood) and nitisinone (69 μmol/l at birth) levels in the newborn; § date of birth

the neurological development up till 12 months of age can be considered normal.

Discussion and Conclusions

We describe an adult female tyrosinaemia type I patient, who unintentionally became pregnant during adequate intake of oral contraceptive. She did not miss any pills, was not ill, did not experience any gastroenteritis episode with vomiting and/or diarrhea before becoming pregnant. There are no data on the metabolization of nitisinone and possible interactions with the metabolism of oral contra-ceptives, so the possibility of a decreased efficacy of her oral contraceptive cannot be formally excluded.

As the patient was already 27 weeks pregnant when the pregnancy was detected, and as the fetal ultrasound and MRI were normal, it was decided that continuation of the treatment (diet, amino acid mixture, and nitisinone) was preferable to changing the treatment or cessation of nitisinone, which would create an unpredictable risk. This decision was thoroughly discussed with all involved parties. During pregnancy, no treatment changes were made, except for an increase in intake of amino acid mixture and natural protein. This adjustment is comparable to what is custom-ary in case of pregnancy in a phenylketonuria mother, to

keep Phe levels within normal range and supply sufficient protein to allow normal growth of the fetus. In this case, natural protein intake was increased because Phe levels fell below normal values (Fig. 3). Further increase of natural protein intake was decided against, in view of the already high maternal Tyr levels.

Nitisinone in the umbilical cord blood was 69 μmol/l at birth, while in the mother it was 94 μmol/l at delivery (paired sampling, last intake of 10 mg of nitisinone 4 h prior to delivery) (Figs. 3 and 4). This indicates that nitisinone passes the placenta almost freely. Five days after birth, the nitisinone level in the newborn was already below the reference target treatment values and on day 28, no nitisinone could be detected anymore (Fig. 4). There was no detectable succinylacetone in the bloodspot of the newborn at any time point. Half-life of nitisinone in the neonate was calculated to be 90 h.

Phe levels at birth were normal in the fetus (80 μmol/l) and at the lower limit of normal values in the mother (30 μmol/l) (Figs. 3 and 4). The ratio between maternal and fetal Phe is comparable to what is described in pregnancies in phenylketonuria mothers and reflects active transplacental transport of Phe. Fetal plasma Phe is on average 1.5 times higher than maternal Phe (Schoonheyt et al. 1994). Tyr level in the neonate at birth was 1,157 μmol/l, while in the mother it was 434 μmol/l (Figs. 3 and 4). The level of Tyr in the neonate probably reflects the fact that nitisinone also inhibits Tyr metabolism in the fetus, but could also partly be due to active transplacental transport. Indeed, fetal plasma Tyr is on average 2.2 ± 0.5 (mean ± 1 standard deviation) times that of the mother (Mitchell et al. 2001). In conclusion, in this first ever case of pregnancy under nitisinone therapy, with moderately elevated maternal Tyr and normal to decreased maternal Phe, outcome for the child seems acceptable so far, at the age of 12 months. Long-term results in a larger cohort of pregnancies and births have to be awaited.

Acknowledgments David Cassiman is a fundamental – clinical researcher for FWO-Vlaanderen.

Roland Devlieger is supported by FWO-Vlaanderen.

Karel Allegaert is supported by FWO-Vlaanderen.

A Concise, 1 Sentence Take-Home Message

Normal 1-year development in a child born from a mother with tyrosinaemia type 1 taking nitisinone during pregnancy.

References

Cerone R, Fantasia AR, Castellano E, Moresco L, Schiaffino MC, Gatti R (2002) Pregnancy and tyrosinaemia type II, case report. J inherit Metab Dis 25:317–318

Davit-Spraul A, Rhomdame H, Poggi-Bach J (2011) Simple and fast quantification of nitisone (NTBC) using liquid chromatography tandem mass spectrometry method in plasma of tyrosinemia type 1 patients. J Chromatogr Sci (in press).

Devlieger H, Martens G, Bekaert A, Eeckels R, Vlietinck R (1996) Perinatale activiteiten in Vlaanderen. 94–116 http://www.neo-natologie.ugent.be/SPE-standaarden.pdf. Accessed 31 May 2011

Francis DE, Kirby DM, Thompson GN (1992) Maternal tyrosinaemia II: management and successful outcome. Eur J Pediatr 151:196–9

Grompe M (2001) The pathophysiology and treatment of hereditary tyrosinemia type 1. Semin Liver Dis 21:563–571

Kvittingen EA, Holme E (2000) Disorder of tyrosine metabolism. In: Fernandes J, Saudubray JM, Van den Berghe G (eds) Inborn metabolic diseases, diagnosis and treatment. Springer, Germany, pp 186–194

Mitchell GA, Grompe M, Lambert M, Tanguay RM (2001) Hyper-tyrosinemia, in Scriver CR, Beadudet AL, Sly WS, Valle D (eds) The Metabolic & Molecular Bases of Inherited Disease, 8th Ed, MCcGraw-Hill, New York, p 1778

Schoonheyt WE, Clarke JT, Hanley WB, Johnson JM, Lehotay DC (1994) Feto-maternal plasma phenylalanine concentration gradient from 19 weeks gestation to term. Clin Chim Acta 225:165–9

Scott CR (2006) The genetic tyrosinemias. Am J Med Genet C 142C:121–126

JIMD Reports
DOI 10.1007/8904_2011_89

RESEARCH REPORT

Molybdenum Cofactor Deficiency: A New HPLC Method for Fast Quantification of S-Sulfocysteine in Urine and Serum

Abdel Ali Belaidi · Sita Arjune ·
Jose Angel Santamaria-Araujo · Jörn Oliver Sass ·
Guenter Schwarz

Received: 25 April 2011 / Revised: 20 July 2011 / Accepted: 01 September 2011 / Published online: 17 December 2011
© SSIEM and Springer-Verlag Berlin Heidelberg 2011

Abstract Molybdenum cofactor deficiency (MoCD) is a rare inherited metabolic disorder characterized by severe and progressive neurological damage mainly caused by the loss of sulfite oxidase activity. Elevated urinary levels of sulfite, thiosulfate, and S-sulfocysteine (SSC) are hallmarks in the diagnosis of MoCD and sulfite oxidase deficiency (SOD). Recently, a first successful treatment of a human MoCD type A patient based on a substitution therapy with the molybdenum cofactor precursor cPMP has been reported, resulting in nearly complete normalization of MoCD biomarkers. Knowing the rapid progression of the disease symptoms in nontreated patients, an early diagnosis of MoCD as well as a sensitive method to monitor daily changes in SSC levels, a key marker of sulfite toxicity, is crucial for treatment outcome. Here, we describe a fast and sensitive method for the analysis of SSC in human urine samples using high performance liquid chromatography (HPLC). The analysis is based on precolumn derivatization with O-phthaldialdehyde (OPA) and separation on a C18 reverse phase column coupled to UV detection. The method was extended to human serum analysis and no interference with endogenous amino acids was found. Finally, SSC values from 45 pediatric urine, 75 adult urine, and 24 serum samples from control individuals as well as MoCD patients are reported. Our method represents a cost-effective technique for routine diagnosis of MoCD and SOD, and can be used also to monitor treatment efficiency in those sulfite toxicity disorders on a daily basis.

Abbreviations

HPLC	High performance liquid chromatography
MoCD	Molybdenum cofactor deficiency
Moco	Molybdenum cofactor
OPA	O-phthaldialdehyde
SOD	Sulfite oxidase deficiency
SSC	S-sulfocysteine

Communicated by: Verena Peters.

Competing interests: None declared.

A.A. Belaidi · S. Arjune · J.A. Santamaria-Araujo ·
G. Schwarz (✉)
Institute of Biochemistry, Department of Chemistry and Center for Molecular Medicine Cologne, University of Cologne, Zuelpicher Str. 47, 50674, Cologne, Germany
e-mail: gschwarz@uni-koeln.de

J.O. Sass
Labor für Klinische Biochemie & Stoffwechsel, Zentrum für Kinder- und Jugendmedizin, Universitätsklinikum Freiburg, Mathildenstr. 1, 79106, Freiburg, Germany

Introduction

MoCD is a rare inherited metabolic disorder (Johnson et al. 1980; Johnson and Duran 2001) caused by defects in the biosynthesis of the molybdenum cofactor (Moco) leading to the simultaneous loss of activities of all molybdenum-dependent enzymes: sulfite oxidase, xanthine dehydrogenase, aldehyde oxidase and mitochondrial amidoxime-reducing component (mARC) (Schwarz et al. 2009). Affected patients exhibit severe neurological abnormalities, such as microcephaly, seizures, and usually die in early childhood (Johnson and Duran 2001). Sulfite oxidase deficiency (SOD) is less frequent but clinically similar to MoCD, which renders sulfite oxidase as the most important

Moco enzyme in humans (Tan et al. 2005). Sulfite oxidase catalyzes the oxidation of sulfite, which is generated throughout the catabolism of sulfur-containing amino acids, to sulfate (Griffith 1987; Johnson and Duran 2001). Xanthine dehydrogenase, another important Moco enzyme, catalyzes the conversion of hypoxanthine to xanthine and further to uric acid (Hille 2005). Aldehyde oxidase catalyzes the hydroxylation of various compounds (Garattini et al. 2009), while mARC was identified as a general detoxifying and pro-drug metabolizing enzyme with yet unknown physiological substrates (Wahl et al. 2010).

Deficiencies of Moco and sulfite oxidase result in the accumulation of sulfite, a highly toxic molecule that breaks disulfide bridges in proteins and cystine, thereby affecting many protein and cellular functions (Zhang et al. 2004). Sulfite first accumulates in liver, where most of the catabolism of sulfur-containing amino acids takes place. Subsequently, accumulation of sulfite in plasma is detectable and finally sulfite crosses the blood–brain-barrier triggering a devastating and progressive neuronal death (Schwarz et al. 2009). Sulfite accumulation is accompanied by the formation of secondary metabolites such as thiosulfate and S-sulfocysteine (SSC) which are common biochemical indicators for MoCD and SOD together with reduced homocysteine levels (Johnson and Duran 2001; Sass et al. 2004). However, isolated SOD can be excluded if the products of other Moco enzymes such as uric acid are reduced in urine and plasma and urinary xanthine and hypoxanthine are elevated. In addition, the absence of urinary urothione as a specific metabolic degradation product of Moco, is a direct indicator for MoCD (Bamforth et al. 1990).

MoCD can be grouped into three types according to the underlying genetic defect. Type A deficiency affects two-thirds of all patients and is caused by a mutation in the *MOCS1* gene (Reiss and Johnson 2003). While type A patients lack the first precursor in the biosynthetic pathway of Moco, cyclic pyranopterin monophosphate (cPMP) (Santamaria-Araujo et al. 2004), type B patients accumulate cPMP due to mutations in the *MOCS2* gene, which encodes the molybdopterin synthase (Reiss et al. 1999). Up to now only one type C patient with a mutation in the *GEPHYRIN* gene has been found (Reiss et al. 2001). Using an animal model for human MoCD (Lee et al. 2002), a treatment approach based on a substitution therapy with cPMP has been established for MoCD type A (Schwarz et al. 2004). Recently, a first human exposure to cPMP has been reported in a type A patient with a remarkable normalization of MoCD biomarkers such as sulfite, SSC, xanthine, uric acid, and urothione leading to a significant clinical improvement of the patient (Veldman et al. 2010). This study revealed also the devastating character of the disease, which was manifested by a rapid increase of urinary sulfite, thiosulfate, and SSC values during the first 36 days of life before treatment started. A recent study supported previous data from animal studies and demonstrated that maternal sulfite clearance is able to suppress prenatal brain damage; however, within days after birth a rapid and progressive brain damage with repetitive seizures has occurred (Sie et al. 2010). Therefore, treatment should be initiated as early as possible. Consequently, the need for a fast and sensitive method for an easy diagnosis of MoCD and a daily monitoring of treatment efficiency becomes important.

In the past, reports showed that diagnosis of MoCD is often difficult, as sulfite is very unstable and false-negative results can occur already within four hours of storage of urine (Kutter and Humbel 1969). False-positive results may also occur in the presence of drugs containing reactive sulfhydryl groups such as *N*-acetylcysteine, mercaptamine, and dimercaprol (Wadman et al. 1983). In contrast, SSC is more stable and suitable for automated analysis and screening. Various methods for determination of SSC were reported, including HPLC-based techniques, electrospray tandem mass spectrometry, and anion exchange chromatography (Demarco et al. 1965; Johnson and Rajagopalan 1995; Pitt et al. 2002). However, only some of these methods provide sufficient sensitivity for the determination of SSC in urine samples, while extended analysis time, complex sample clean-up procedures, or instability of certain SSC derivatives limited their use in an automated, fast, and reproducible setting. A general method currently used for the diagnosis of MoCD is based on a sulfite dipstick for the rapid detection of urinary sulfite. Furthermore, most SSC analyses are performed with the classical amino acid analyzer, but quantification is limited due to the early elution of SSC. In addition, poor specificity and possible interference from other metabolites were also reported by using the amino acid analyzer (Rashed et al. 2005).

Here, we describe a simple, fast, and sensitive HPLC-based method using automated precolumn derivatization with OPA and UV detection at 338 nm for the determination of urinary SSC levels. The method was extended to serum analysis and SSC separation from endogenous amino acids was demonstrated. We reported SSC control levels derived from 45 pediatric urine, 75 adult urine, and 24 adult serum samples. Furthermore, SSC values from MoCD patients are reported.

Materials and Methods

Reagents

HPLC-grade methanol, acetonitrile, and disodium hydrogen phosphate were obtained from BDH Prolabo

(VWR International GmbH, Darmstadt, Germany). Sodium tetraborate, boric acid, 5-Sulfosalicylic acid, and *S*-sulfocysteine were from Sigma–Aldrich (St. Louis, USA). *O*-Phthaldialdehyde (OPA) was purchased from Alfa Aesar GmbH & Co KG (Karlsruhe, Germany). Preparation of the derivatization reagent was based on the protocol of Jakoby (Jakoby and Griffith 1987) and achieved by dissolving 1 g of OPA in 10 ml methanol, mixed with 90 ml 0.4 M borate buffer pH 10.2 and 400 µl 2-mercaptoethanol. The prepared derivatization reagent was stored in 1 ml aliquots at −20°C.

Chromatographic Conditions

HPLC analyses were carried out on an Agilent 1200 SL system (Agilent Technologies GmbH, Boeblingen, Germany) consisting of a binary pump (SL series), vacuum degasser, autosampler, thermostated column compartment (SL series), diode array detector (SL series), fluorescence detector, all controlled by Agilent ChemStation software. The mobile phase consisted of buffer A (10 mM Na_2HPO_4; 10 mM $Na_2B_4O_7$) and eluent B (45% acetonitrile 45% methanol and 10% water). Different reversed-phase BEH (ethylene bridged hybrid)-C18 columns were used: XBridge (50 × 4.6 mm, 2.5 µm, Waters GmbH, Eschborn, Germany) and ACQUITY UPLC (50 × 2.1 mm, 1.7 µm, Waters GmbH, Eschborn, Germany) and separation was carried out isocratically with 10% eluent B. For analysis of all amino acids, an additional gradient from 10 to 60% eluent B was applied to the column after elution of SSC. Precolumn derivatization was carried out using an automated autosampler, which was programmed to mix 9 µl of the sample with 1 µl of derivatization reagent, after an incubation time of 0.2 min, the mixture was injected and the injection valve was bypassed to achieve the derivatization of the next sample, thus eliminating the time needed for the derivatization procedure. Detection was carried out by UV absorbance at 338 nm and compound identification was achieved by comparing the retention time with that obtained for SSC standard. Peak area was used for calibration. SSC amount was determined by standard addition and normalized to creatinine in urine samples.

Samples and Standard Preparation

Urine samples were centrifuged for 15 min at 13,000 × g, the supernatant was filtered through a 0.2-µm filter and no additional solid phase extraction was performed. Urine samples were diluted in HPLC-grade water and directly analyzed. For the analysis of serum samples, fresh blood samples were centrifuged for 10 min at 1,300 × g and serum was collected and stored at −20°C until further analysis. For serum protein precipitation, 100 µl 5% 5-sulfosalicylic acid were added to 200 µl serum and 200 µl

HPLC-grade water. Precipitated proteins were removed by centrifugation for 15 min at 13,000 × g. The supernatant fraction was filtered through a 0.2-µm filter and directly analyzed.

Creatinine Analysis

Creatinine determination was based on the Jaffe method (Vasiliades 1976; Kroll et al. 1984). Briefly, 50 µl of diluted urine samples were mixed with 150 µl alkaline picrate solution (1.2% picric acid in 0.75 M sodium hydroxide). After an incubation time of 15 min, the formation of an orange-red complex between creatinine and alkaline picrate was quantified by measuring the absorbance at 490 nm using a microplate reader (BioTek, Friedrichshall, Germany). Method calibration was achieved by measuring the absorbance of different creatinine standard solutions.

Validation Procedure: Linearity and Calibration

Calibration of the method was carried out using three different conditions. SSC standard solutions were added to water, SSC-free urine, and serum samples. Peak area was used for quantification and linearity was determined by using a wide SSC concentration range from 2.5 to 200 µM.

Validation Procedure: Specificity

SSC determination was carried out using isocratic elution from the HPLC column. Under those conditions, the majority of the amino acids were eluted after the SSC peak. To demonstrate this, a standard solution containing all primary amino acids including SSC and taurine was separated. The specificity of the method was also determined by analyzing urine and serum samples under the same chromatographic conditions to show separation of SSC from endogenous amino acids.

Validation Procedure: Limit of Detection and Limit of Quantification

Limit of detection (LOD) and limit of quantification (LOQ) were determined by measuring the standard deviation of peak area and slope of the calibration curve. LOD and LOQ were determined by adding increasing amounts of SSC standard in the range of 2.5–200 µM to water, SSC-free urine, and serum samples.

Validation Procedure: Precision and Reproducibility

Intraday precision was determined by analyzing samples three times a day randomly and interday precision was determined by analyzing samples at ten different days.

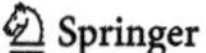

Relative standard deviation (RSD) of peak area was used and analysis was carried out at three different concentrations. Low quantity control (LQC), median quantity control (MQC), and high quantity control (HQC) samples were prepared by spiking, water, SSC-free urine, and serum samples to a final concentration of 10, 60, and 200 µM, respectively. Reproducibility of the method was assayed by using different matrix batches, different column diameters, and lengths. Different separation methods were tested including isocratic or gradient elution at different flow rates.

Validation Procedure: Recovery

The percentage of recovery was determined at three concentration levels LQC, MQC, and HQC. SSC-free urine and serum samples were spiked with SSC standard. Peak area of the response was compared to that of standard solutions prepared in water.

Results and Discussion

Derivatization

HPLC analysis of amino acids with OPA derivatization is one of the most sensitive methods for amino acids quantification with detection limits in the femtomole range. OPA reacts with all primary amino acids under alkaline conditions and in the presence of thiols OPA forms isoindoles that can be detected either by fluorescence (excitation 230 nm, emission 450 nm) or absorbance at 340 nm (Hill et al. 1979). Amino acids carrying a free thiol group such as cysteine or homocysteine cannot be directly detected with OPA derivatization and a modification prior to the derivatization reaction is needed. Carboxymethylation is usually used to convert cysteine to carboxymethylcysteine prior to OPA derivatization, thus leading to a fluorescence intensity similar to other amino acids (Jarrett et al. 1986). Based on the chemical structure of SSC lacking a free thiol group, we have chosen to use a direct derivatization of SSC with OPA without prior modification.

Chromatography

One of the critical steps in the OPA derivatization procedure is the instability of the amino acid derivatives, which is manifested by the decay of fluorescence over time (Hogan et al. 1982; Fleury and Ashley 1983; Cooper et al. 1984). To overcome this problem, we aimed to develop fast chromatographic methods using short columns (2.1 mm × 50 mm) and UPLC particles (reversed phase C18 1.7 µm) to achieve high resolution at short analysis time. As a

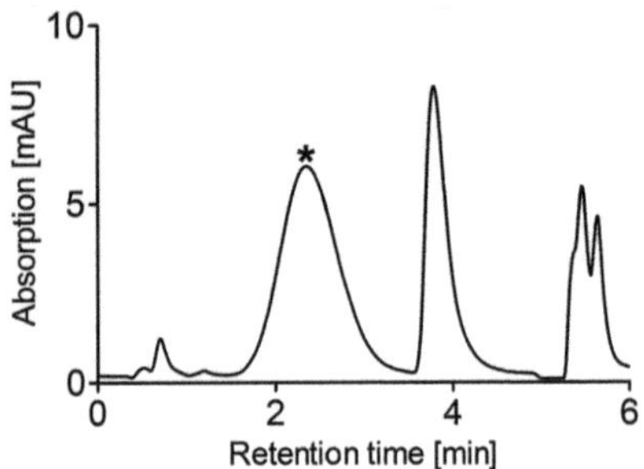

Fig. 1 HPLC analysis of standard SSC solution. Twenty micromolar SSC standard was prepared in H$_2$O and analyzed under chromatographic conditions described in Materials and Methods. Retention time of SSC peak was 3.8 min. The large peak eluting before SSC (depicted by an *asterisk*) represents a degradation product of OPA. The OPA excess is used to ensure quantitative derivatization of all primary amino acids in urine and serum

result, separation was completed within 6 min using isocratic elution allowing rapid analysis. Under the chromatographic conditions described, SSC yielded a sharp peak eluting at 3.8 min (Fig. 1). Specificity of the method was determined by analyzing a standard solution containing all primary amino acids including SSC and taurine and a gradient elution was applied after elution of SSC. The results showed that most of the amino acids are eluted after the SSC peak, and thus the chromatographic conditions optimized for SSC allowed a clear separation from all other amino acids (Fig. 2a).

Detection of OPA derivatives is carried out by using either fluorescence or UV absorbance. However, SSC showed a very low specific fluorescence when compared to other primary amino acids (Fig. 2b). Furthermore, all primary amino acids and SSC OPA-derivatives showed a similar absorbance ratio at 338 nm; thus, using the UV approach a direct identification of accumulated OPA derivatives can be easily achieved.

We analyzed urine and serum samples derived from control individuals for SSC content and the SSC peak was separated (Fig. 3a, b). The SSC peak was identified by comparing its retention time to that of a separately injected standard peak and peak purity was proven by spectral homogeneity (data not shown). SSC concentration was determined by standard addition (Fig. 3a, b). Analysis of samples derived from healthy individuals revealed the presence of SSC in urine, while in serum SSC was below detection limit. In contrast, analysis of urine and serum samples derived from MoCD patients showed an accumulated SSC peak, which was quantified in both urine and serum samples (Fig. 3c, d).

HPLC analysis of OPA-derivatized amino acids is usually carried out using alkaline buffers, which provide a greater stability of the amino acid derivatives (Cooper et al. 1984). However, a major problem of alkaline running buffers is the

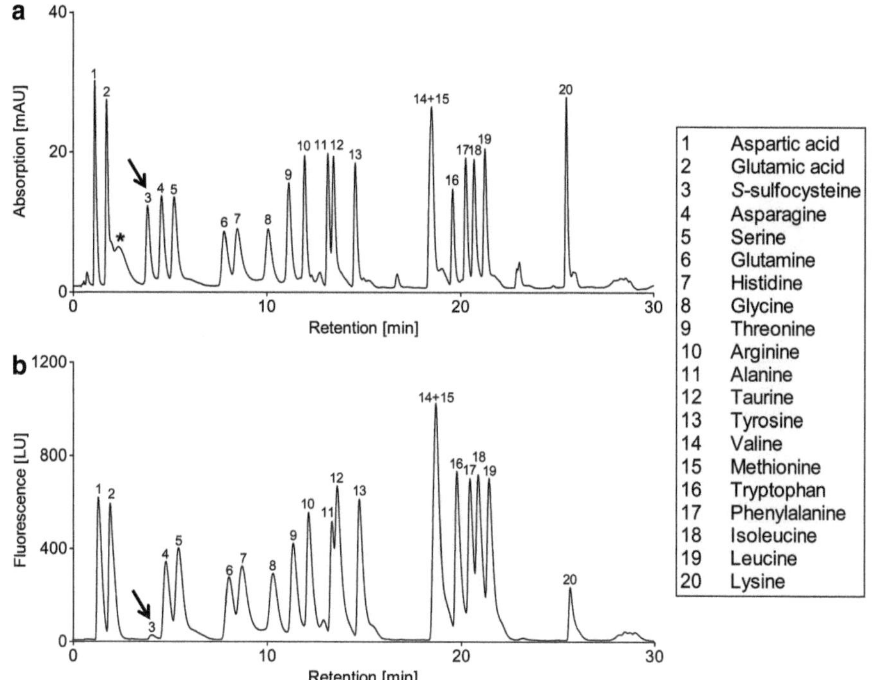

Fig. 2 HPLC analysis of amino acids. A standard solution of 20 μM containing all primary amino acids and additionally SSC and taurine were prepared in H_2O and separated under chromatographic conditions described in Materials and Methods. Detection was carried out by UV absorbance at 338 nm (**a**) and fluorescence (**b**). All amino acids are numbered and shown in the legend. The SSC peak is depicted by an *arrow* in both panels. The differences in peak shape resulted from the different elution methods used: For the first eight amino acids isocratic elution was used while the remaining 12 amino acids were eluted with a gradient from 10 to 60% of eluent B

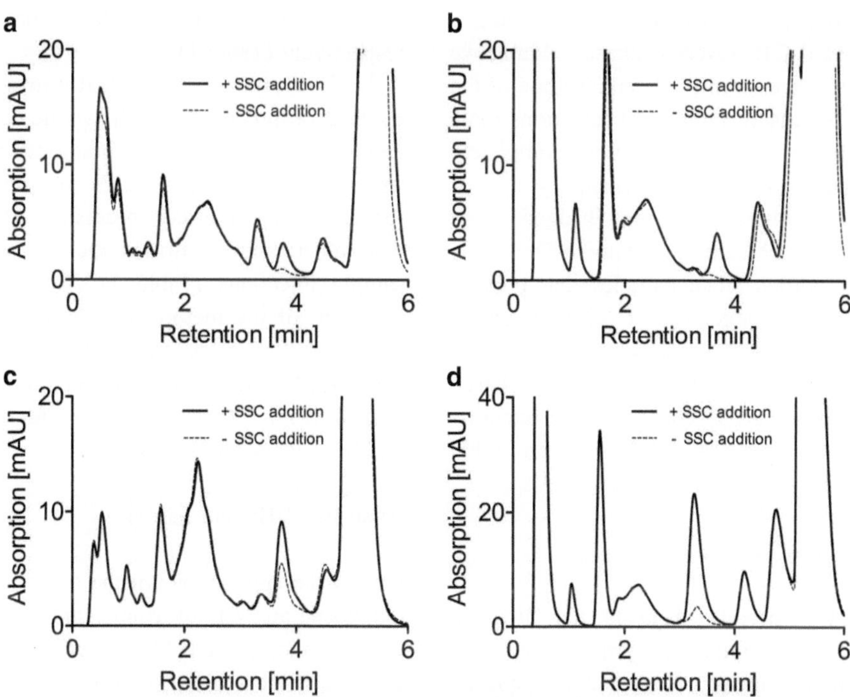

Fig. 3 Standard addition method for quantification of SSC in urine and serum. Control urine (**a**) and control serum (**b**) was analyzed and the standard addition method was used to quantify SSC. Control urine sample shows an SSC peak close to the detection limit whereas in control serum SSC was not detectable. Analysis of urine (**c**) and serum (**d**) samples derived from non-treated MoCD patient show a clear accumulation of SSC. All chromatograms show an overlay of two analyses. The first analysis was carried out by diluting samples with H_2O (−SSC addition) while prior to the second analysis the samples were diluted with SSC standard (+SSC addition). SSC amounts were quantified by comparing peak area in both analyses and SSC amounts were normalized to creatinine for urine samples and expressed in mmol/mol creatinine

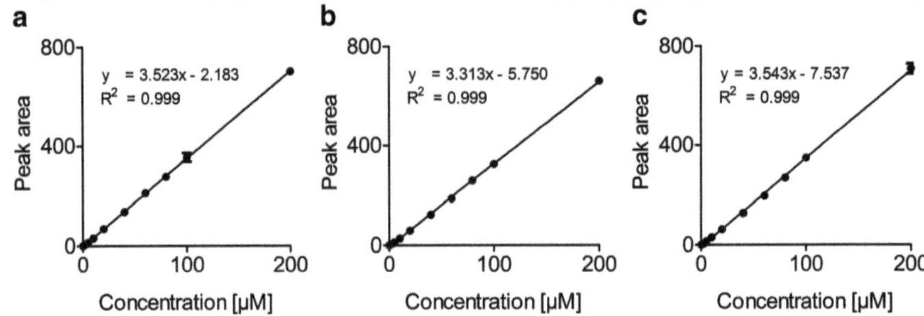

Fig. 4 Liniarity of SSC quantification. Ten SSC concentrations ranging from 0 to 200 µM were spiked in H_2O (**a**), SSC-free urine (**b**) and SSC-free serum (**c**) and subsequently used for method calibration. All samples were run in triplicates and points represent means ± relative standard deviations (RSD) of peak area. Equation and correlation coefficient (R^2) are given for the linear regression curves of the three different conditions

Table 1 Method sensitivity, precision and recovery of SSC

Specimen	LOD (µM)	LOQ (µM)	Intraday stability RSD (%)			Interday stability RSD (%)			Recovery (%)		
			LQC	MQC	HQC	LQC	MQC	HQC	LQC	MQC	HQC
H_2O	1.18	3.58	1.12	3.86	2.19	3.34	4.15	0.57	–	–	–
Urine	1.44	4.38	0.76	1.45	0.81	8.22	1.97	6.65	96.10–97.73	99.60–100.75	96.09–97.48
Serum	1.08	3.27	0.70	1.04	2.77	0.56	3.87	3.40	95.77–96.72	92.16–93.53	98.89–102.84

short lifetime of silica-based columns. The pH of the mobile phase used under our chromatographic conditions was not adjusted and corresponded to pH 9.2. Using a conventional silica-based C18 reversed phase column, we observed a continuous decrease in the retention time of the SSC derivative with increasing number of injections (data not shown). After 500 injections, the column completely lost its separation performance. The sensitivity against alkaline buffers was resolved by using Bridged Ethyl Hybrid (BEH) particles (Waters, Germany). The BEH particles consist of modified silica particles that provide increased pH stability in the range of pH 1–12 compared to the conventional silica-based C18 resins with a maximal suitable pH of 8.0. Using the BEH C18 silica-based columns, we did not observe any change in the retention time of any analyzed derivatives for more than 5,000 injections.

Calibration, Precision, and Recovery

Method calibration was performed using ten different concentrations of SSC standard ranging from 0 to 200 µM. All calibration standards were run in triplicate and peak area was used for quantification. Linearity of the method was determined in three different experimental set-ups. Different concentrations of standard solutions were prepared in (1) water, (2) urine, or (3) serum and subsequently analyzed. The calibration curves displayed high linearity over the concentration range investigated (Fig. 4). LOD ranged from 1.08 to 1.44 µM and LOQ ranged from 3.27 to 4.38 for serum and urine analysis, respectively (Table 1).

Derivatization methods add extra steps in the sample preparation procedure resulting often in variations between repeated injections. By using automated precolumn derivatization these variations can be reduced to a minimum. Intraday and interday precision of the method were determined and confirmed the low variation in repeated sample injections (Table 1). The intraday and interday precision of the method ranged from 0.70 to 3.86% and 0.56 to 8.22%, respectively. Recovery was determined for urine and serum at three concentration levels of SSC and ranged from 87.34 to 94.76% for urine and 92.16 to 102.84% for serum (Table 1).

Analysis of Human Samples

One of the applications of the described method is the fast diagnostic analysis of urine samples derived from individuals with suspected MoCD or SOD. A positive sulfite dipstick is the starting point of diagnostic tests commonly used in clinics and most laboratories. A typical diagnostic procedure usually includes two steps: (1) fresh urine sample is analyzed and SSC peak is identified by comparing the retention time with that of standard and (2) standard addition verifies the identity of the peak and quantifies

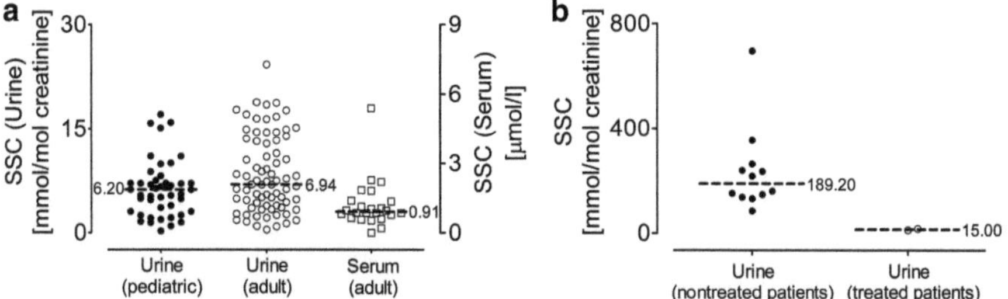

Fig. 5 SSC levels in control individuals and MoCD patients. SSC was quantified in urine and serum samples derived from control individuals (**a**) and urine samples from MoCD patients (**b**). In total, 45 pediatric urine samples (*filled circles*), 75 adult urine (*empty circles*) and 24 adult serum samples (*empty squares*) were analyzed to determine the SSC threshold in control individuals. SSC values of urine samples derived from 12 nontreated MoCD patients are shown in panel **b** (*filled circles*) and from two cPMP-treated MoCD patients (*empty circles*). SSC amount is given in mmol/mol creatinine and µmol/l for urine and serum samples, respectively. Median values are shown by *horizontal bars*

SSC. For this, a defined amount of SSC was added to urine or serum sample, and the resulting chromatogram was compared to the chromatogram obtained without standard addition. A positive control displayed a symmetric SSC Peak with an increased area (Fig. 3c, d).

Here, the method was applied to the analysis of 45 pediatric urine, 75 adult urine, and 24 serum samples all derived from control individuals. SSC was detected in all analyzed urine samples demonstrating the low detection limit of the method (Fig. 5a). The values in pediatric urine samples ranged from 0.26 to 18.83 mmol/mol creatinine (median: 6.20 mmol/mol creatinine), and SSC levels of adult urine samples were comparable with a median of 6.94 mmol/mol creatinine. This finding demonstrates a similar SSC excretion level in both age groups. Serum analysis revealed in general low SSC levels and the values ranged from 0 to 5.39 µmol/l with a median of 0.91 (Fig. 5a). This finding suggests a continuous excretion of SSC in urine without any threshold barrier.

Different urine samples from nontreated MoCD patients were analyzed and SSC was detectable from the first days of life. SSC analysis of urine samples derived from 12 genetically confirmed MoCD type A or type B individuals showed SSC levels ranging from 85 to 695 mmol/mol creatinine (Fig. 5b). Except the index case (Veldman et al. 2010), all currently cPMP-treated MoCD patients (Schwarz et al., unpublished results) are monitored for their SSC levels using the here described method. Here, we show average data over a six months period for two patients treated with cPMP (Schwahn et al. 2010) (Fig. 5b).

Pitt and colleagues reported SSC urinary levels of 0–9 mmol/mol creatinine for control individuals (Pitt et al. 2002) with a median of approximately 2 mmol SSC/mol creatinine, which are lower than the SSC urinary levels measured in this study with a median of 6.94 mmol/mol creatinine. However, the authors reported a poor recovery of SSC in low controls by using MS in negative-ion mode, which they attribute to high salt content of urine samples

and the fact that a single internal standard was used. We did not observe any vulnerability of the method used in this study to the ionic content of the sample. Furthermore, we quantified SSC by the use of an identical internal standard. In addition, the reported SSC urinary values of a cPMP-treated MoCD type A patient (Veldman et al. 2010) as well as the reported values of two cPMP-treated MoCD patients in this study are in the same range as the values determined for healthy individuals. Taking into account that urine analysis was carried out without precleaning procedures, thus maximizing recovery, we conclude that the urinary SSC values presented in this study can serve as reference values, which should help to define a threshold limit for future routine diagnostics.

Conclusion

We present a simple, fast, and robust method for the determination of SSC in human urine. No sample clean-up was required, thus reducing analysis time and the risk of contamination. We have shown that our method can be extended to the determination of serum SSC levels with the inclusion of an additional precipitation step. High sensitivity of the method enabled determination of urinary SSC levels in control pediatric and adult individuals revealing a similar excretion range with a median of 6.20 and 6.94 mmol/mol creatinine, respectively. The method has been successfully used for the diagnosis of MoCD/SOD and routine monitoring of cPMP-treated MoCD patients (Schwarz et al., unpublished results). The sensitivity, linearity, and accuracy of the HPLC derivatization method demonstrated in this study, together with the short analysis time, makes this method suitable for automated diagnostic analysis which can be used in a multi-well plate-based autosampler. Given the fact that MoCD is still considered an underdiagnosed inborn error in metabolism (Johnson and Duran 2001), the described method should provide

diagnostic laboratories with an accurate, fast, and cost-effective method to quantify SSC levels in urine and serum.

Acknowledgments We thank Parmanand Arjune, MD, PhD (Cologne, Germany) for providing a large number of urine and serum samples. We are grateful to James Pitt, PhD (Melbourne, Australia) for the comparative analysis of selected urine samples. We gratefully acknowledge Bernd Schwahn, MD, Steve Bowhay, PhD, Peter G. Galloway, MD (Royal Hospital for Sick Children, Glasgow, Scotland), Julia Hennemann, MD (Pediatric Unit, Charité Berlin, Germany), Francjan J. van Spronsen, MD, PhD (University Medical Center, Groningen, Netherlands), and Illona Weis, MD (Pediatric Unit, Gemeinschaftsklinikum Koblenz-Mayen, Germany) and other colleagues for providing urine samples of patients.

References

Bamforth FJ, Johnson JL, Davidson AG, Wong LT, Lockitch G, Applegarth DA (1990) Biochemical investigation of a child with molybdenum cofactor deficiency. Clin Biochem 23:537–542

Cooper JD, Ogden G, McIntosh J, Turnell DC (1984) The stability of the o-phthalaldehyde/2-mercaptoethanol derivatives of amino acids: an investigation using high-pressure liquid chromatography with a precolumn derivatization technique. Anal Biochem 142:98–102

Demarco C, Mosti R, Cavallin D (1965) Column chromatography of some sulfur-containing amino acids. J Chromatogr 18:492

Fleury MO, Ashley DV (1983) High-performance liquid chromatographic analysis of amino acids in physiological fluids: on-line precolumn derivatization with o-phthaldialdehyde. Anal Biochem 133:330–335

Garattini E, Fratelli M, Terao M (2009) The mammalian aldehyde oxidase gene family. Hum Genomics 4:119–130

Griffith OW (1987) Mammalian sulfur amino acid metabolism: an overview. Methods Enzymol 143:366–376

Hill DW, Walters FH, Wilson TD, Stuart JD (1979) High performance liquid chromatographic determination of amino acids in the picomole range. Anal Chem 51:1338–1341

Hille R (2005) Molybdenum-containing hydroxylases. Arch Biochem Biophys 433:107–116

Hogan DL, Kraemer KL, Isenberg JI (1982) The use of high-performance liquid chromatography for quantitation of plasma amino acids in man. Anal Biochem 127:17–24

Jakoby WBE, Griffith OWE (1987) Sulfur and sulfur amino acids. Academic, London

Jarrett HW, Cooksy KD, Ellis B, Anderson JM (1986) The separation of o-phthalaldehyde derivatives of amino acids by reversed-phase chromatography on octylsilica columns. Anal Biochem 153:189–198

Johnson JL, Duran M (2001) Molybdenum cofactor deficiency and isolated sulfite oxidase deficiency. In: Scriver C et al (eds) The metabolic and molecular bases of inherited disease. McGraw-Hill, New York, pp 3163–3177

Johnson JL, Rajagopalan KV (1995) An HPLC assay for detection of elevated urinary S-sulphocysteine, a metabolic marker of sulfite oxidase deficiency. J Inherit Metab Dis 18:40–47

Johnson JL, Waud WR, Rajagopalan KV, Duran M, Beemer FA, Wadman SK (1980) Inborn errors of molybdenum metabolism: combined deficiencies of sulfite oxidase and xanthine dehydrogenase in a patient lacking the molybdenum cofactor. Proc Natl Acad Sci U S A 77:3715–3719

Kroll MH, Hagengruber C, Elin RJ (1984) Reaction of picrate with creatinine and cepha antibiotics. Clin Chem 30:1664–1666

Kutter D, Humbel R (1969) Screening for sulfite oxidase deficiency. Clin Chim Acta 24:211–214

Lee HJ, Adham IM, Schwarz G, Kneussel M, Sass JO, Engel W, Reiss J (2002) Molybdenum cofactor-deficient mice resemble the phenotype of human patients. Hum Mol Genet 11:3309–3317

Pitt JJ, Eggington M, Kahler SG (2002) Comprehensive screening of urine samples for inborn errors of metabolism by electrospray tandem mass spectrometry. Clin Chem 48:1970–1980

Rashed MS, Saadallah AA, Rahbeeni Z, Eyaid W, Seidahmed MZ, Al-Shahwan S, Salih MA, Osman ME, Al-Amoudi M, Al-Ahaidib L, Jacob M (2005) Determination of urinary S-sulphocysteine, xanthine and hypoxanthine by liquid chromatography-electrospray tandem mass spectrometry. Biomed Chromatogr 19:223–230

Reiss J, Johnson JL (2003) Mutations in the molybdenum cofactor biosynthetic genes MOCS1, MOCS2, and GEPH. Hum Mutat 21:569–576

Reiss J, Dorche C, Stallmeyer B, Mendel RR, Cohen N, Zabot MT (1999) Human molybdopterin synthase gene: genomic structure and mutations in molybdenum cofactor deficiency type B. Am J Hum Genet 64:706–711

Reiss J, Gross-Hardt S, Christensen E, Schmidt P, Mendel RR, Schwarz G (2001) A mutation in the gene for the neurotransmitter receptor-clustering protein gephyrin causes a novel form of molybdenum cofactor deficiency. Am J Hum Genet 68:208–213

Santamaria-Araujo JA, Fischer B, Otte T, Nimtz M, Mendel RR, Wray V, Schwarz G (2004) The tetrahydropyranopterin structure of the sulfur-free and metal-free molybdenum cofactor precursor. J Biol Chem 279:15994–15999

Sass JO, Nakanishi T, Sato T, Shimizu A (2004) New approaches towards laboratory diagnosis of isolated sulphite oxidase deficiency. Ann Clin Biochem 41:157–159

Schwahn BC, Galloway PG, Bowhay S, Veldman A, Santamaria JA, Schwarz G, Belaidi AA (2010) Successful treatment of two neonates with molybdenum cofactor deficiency (MOCD) type a, using cyclic pyranopterine monophosphate (CPMP). J Inherit Metab Dis 33:S29

Schwarz G, Santamaria-Araujo JA, Wolf S, Lee HJ, Adham IM, Grone HJ, Schwegler H, Sass JO, Otte T, Hanzelmann P, Mendel RR, Engel W, Reiss J (2004) Rescue of lethal molybdenum cofactor deficiency by a biosynthetic precursor from *Escherichia coli*. Hum Mol Genet 13:1249–1255

Schwarz G, Mendel RR, Ribbe MW (2009) Molybdenum cofactors, enzymes and pathways. Nature 460:839–847

Sie SD, de Jonge RC, Blom HJ, Mulder MF, Reiss J, Vermeulen RJ, Peeters-Scholte CM (2010) Chronological changes of the amplitude-integrated EEG in a neonate with molybdenum cofactor deficiency. J Inherit Metab Dis DOI: 10.1007/s10545-010-9198-z (Epub ahead of print).

Tan WH, Eichler FS, Hoda S, Lee MS, Baris H, Hanley CA, Grant PE, Krishnamoorthy KS, Shih VE (2005) Isolated sulfite oxidase deficiency: a case report with a novel mutation and review of the literature. Pediatrics 116:757–766

Vasiliades J (1976) Reaction of alkaline sodium picrate with creatinine: I. Kinetics and mechanism of formation of the mono-creatinine picric acid complex. Clin Chem 22:1664–1671

Veldman A, Santamaria-Araujo JA, Sollazzo S, Pitt J, Gianello R, Yaplito-Lee J, Wong F, Ramsden CA, Reiss J, Cook I, Fairweather J, Schwarz G (2010) Successful treatment of molybdenum cofactor deficiency type A with cPMP. Pediatrics 125: e1249–e1254

Wadman SK, Cats BP, Debree PK (1983) Sulfite oxidase deficiency and the detection of urinary sulfite. Eur J Pediatr 141:62–63

Wahl B, Reichmann D, Niks D, Krompholz N, Havemeyer A, Clement B, Messerschmidt T, Rothkegel M, Biester H, Hille R, Mendel RR, Bittner F (2010) Biochemical and spectroscopic characterization of the human mitochondrial amidoxime reducing components hmARC-1 and hmARC-2 suggests the existence of a new molybdenum enzyme family in eukaryotes. J Biol Chem 285:37847–37859

Zhang X, Vincent AS, Halliwell B, Wong KP (2004) A mechanism of sulfite neurotoxicity: direct inhibition of glutamate dehydrogenase. J Biol Chem 279:43035–43045

JIMD Reports
DOI 10.1007/8904_2011_92

Adenine Phosphoribosyltransferase Deficiency: An Underdiagnosed Cause of Lithiasis and Renal Failure

Giuseppina Marra · Paolo Gilles Vercelloni · Alberto Edefonti ·
Gianantonio Manzoni · Maria Angela Pavesi · Giovanni Battista Fogazzi ·
Giuseppe Garigali · Lionel Mockel · Irene Ceballos Picot

Received: 26 July 2011 / Revised: 5 September 2011 / Accepted: 7 September 2011 / Published online: 21 December 2011
© SSIEM and Springer-Verlag Berlin Heidelberg 2011

Abstract We describe an infant affected by adenine phosphoribosyltransferase (APRT) deficiency diagnosed at 18 months of age with a de novo mutation that has not been previously reported. APRT deficiency is a rare defect of uric acid catabolism that leads to the accumulation of 2,8 dihydroxyadenine (2,8-DHA), a highly insoluble substance excreted by the kidneys that may precipitate in urine and form stones. The child suffered from renal colic due to a stone found in the peno-scrotal junction of the bulbar urethra. Stone spectrophotometric analysis allowed us to diagnose the disease and start kidney-saving therapy in order to avoid irreversible chronic kidney damage. APRT deficiency should always be considered in the differential diagnosis of pediatric urolithiasis.

The authors Giuseppina Marra and Paolo Gilles Vercelloni equally contributed to the article.

Communicated by: Jean-Marie Saudubray.

Competing interests: None declared.

G. Marra (✉) · P.G. Vercelloni · A. Edefonti
Fondazione IRCCS Ca' Granda Ospedale Maggiore Policlinico,
UO Pediatric Nephrology, Via Commenda 9, 20122, Milano, Italy
e-mail: gimarra@hotmail.com

G. Manzoni
Fondazione IRCCS Ca' Granda Ospedale Maggiore Policlinico,
UO Pediatric Surgery, Via Commenda 9, 20122, Milano, Italy

M.A. Pavesi
Fondazione IRCCS Ca' Granda Ospedale Maggiore Policlinico,
UO Pediatric Radiology, Via Commenda 920122, Milano, Italy

G.B. Fogazzi · G. Garigali
Fondazione IRCCS Ca' Granda Ospedale Maggiore Policlinico,
UO Nephrology, Via Commenda 9, 20122, Milano, Italy

L. Mockel · I.C. Picot
Laboratory of Metabolic Biochemistry, Necker-Enfants Malades
Hospital, Assistance Publique-Hôpitaux de Paris, Rue de Sevres,
Paris, France

Introduction

APRT deficiency is a rare defect of uric acid catabolism that leads to the accumulation of 2,8 dihydroxiadenine (2,8-DHA), a highly insoluble substance excreted by the kidney that may precipitate in urine (Bouzidi et al. 2007; Simmonds 2003). The disease is due to a mutation in the APRT gene and is inherited in an autosomal recessive manner (Tischfield and Ruddle 1974). The small number of patients diagnosed in contrast with the mutation frequency in general population suggests that it is a still under-diagnosed disease (Ceballos-Picot et al. 1992). It is not only an uncommon cause of renal stones but can also cause chronic renal failure and lead to dialysis as other lithogenic diseases such as cistinuria and primary hyperoxaluria. Furthermore, unless promptly diagnosed and treated, it may even recur in a transplanted kidney (Benedetto et al. 2001; Bollée et al. 2010; Edvardsson et al. 2001; Stratta et al. 2010). Most of the published case reports concern adults (Simmonds 2003), but it is very important to recognize and treat the disease as soon as possible in order to reduce lithiasic episodes and prevent its potential long-term consequences. A correct diagnosis can be made by following a diagnostic workup but a high index of suspicion is necessary.

We here describe a patient with APRT deficiency who was diagnosed during infancy and was due to a de novo mutation that has not been previously reported.

Case Report

AP was born in the 29th week of gestation by cesarean delivery due to placenta praevia in an uneventful twin pregnancy. The neonatal period was characterized by cardiorespiratory distress and jaundice, which required ICU admission and CPAP and phototherapy. A number of perinatal renal ultrasound (US) examinations revealed micropapillary calcifications probably related to his prematurity without any urinary tract dilatation or other alteration of the renal parenchyma (Fig. 1a). His twin sister had a similar clinical history and showed the same renal findings, which disappeared spontaneously during the first few months of life. In our patient, they remained unchanged (Fig. 1b) and subsequently assumed the appearance of nephrocalcinosis.

At the age of 8 months, he developed a symptomatic upper urinary tract infection. When he was 16 months old, a renal US examination revealed multiple bilateral radiolucent apico-papillary and peripheral stones with diameters ranging from 2 to 5 mm. Two months later, he suffered an acute episode of renal colic, after which two small stones were found in a nappy and a further two were observed by US in the pelvis of both kidneys. Two weeks later, he presented acute urine retention with stranguria and oliguria, and a stone was palpable at the peno-scrotal junction of the bulbar urethra. He underwent an emergency endoscopic

evaluation with stone fragmentation and lithotripsy, and was eventually referred to our hospital for nephrological and metabolic evaluation.

The hematological (acid-base balance, parathormone, electrolytemia, renal function) and urinary examinations showed no signs of tubulopathy or major metabolic diseases causing urolithiasis: aminoaciduria and cystinuria, ossaluria and citraturia were normal; calciuria/creatininuria was 0.27 (95th percentile = 0.56), phosphaturia/creatininuria 2.13 (95th percentile = 3.95), magnesiuria/creatininuria 0.19 (95th percentile = 0.37), and uricuria/creatininuria 0.96 (normal value <1).

A semiqualitative spectrophotometric analysis of the removed stone revealed a curve very similar to that of 2,8-DHA (Fig. 2), adenine phosphoribosyltransferase (APRT) deficiency was suspected, and it was found that the patient's red blood cells did not show any APRT activity. Genetic analysis of the APRT gene confirmed the diagnosis of APRT deficiency, and showed that the patient was composite heterozygous for two mutations: a known mutation c.84C>A localized on exon 2 and leading to the substitution p.Asp28Glu at protein level, and a never previously described deletion (g.347_372del26) located at the exon2/intron2 junction.

In the meantime, medical therapy was started with allopurinol 10 mg/kg day combined with hydropinic therapy.

The child is now 3 years old and has normal renal function. The micropapillary deposits disappeared 1 year after the beginning of therapy (Fig. 1c); the residual pelvic stone is unchanged in size (<5 mm), and no new 2,8-DHA

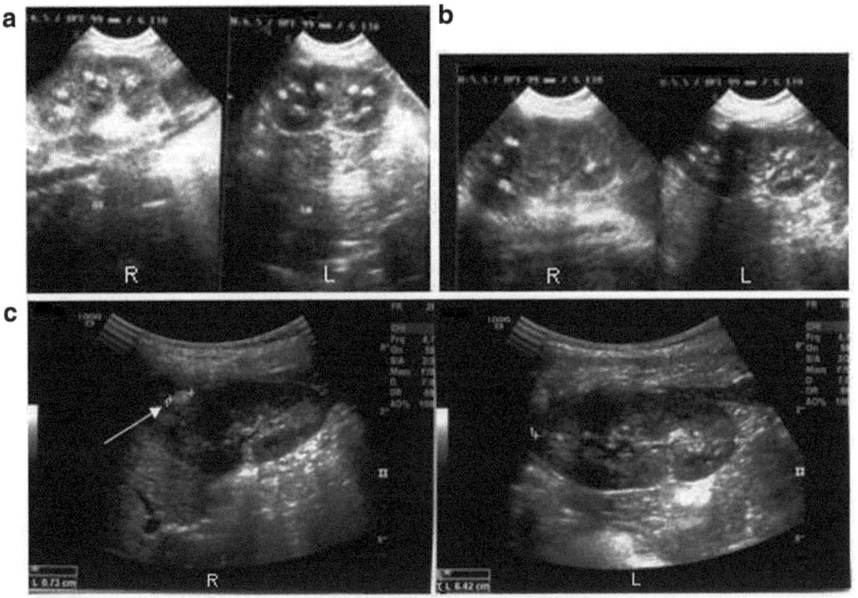

Fig. 1 Ultrasound images before therapy at 3 (**a**) and 9 months of age (**b**), and after 1 year of therapy (**c**)

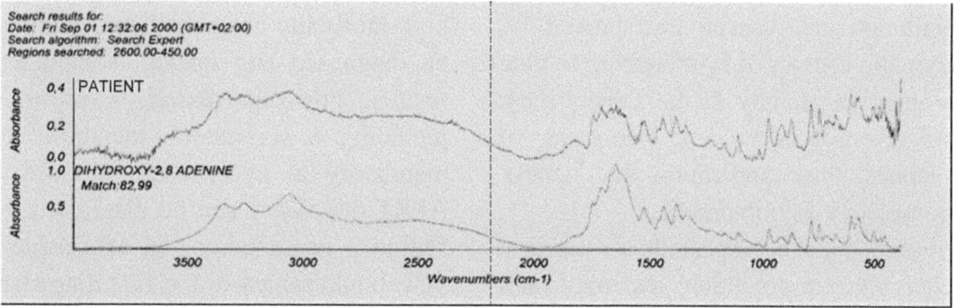

Fig. 2 Spectrophotometry of the surgically extracted stone

stones have formed. Furthermore, no 2,8-DHA crystals were ever found in the many follow-up searches of urine sediment.

After the initial diagnosis was established, the whole family underwent a complete APRT enzymatic and genetic evaluation. The father and twin sister of the index case had a low residual erythrocyte APRT activity (7% and 38%, respectively). Both of them had the c.84C>A mutation heterozygously but not the deletion and various renal US examinations revealed that they both had normal kidneys and urinary tract and no crystalluria. The APRT activity of the index case's mother was normal and no genetic abnormalities were found. This familial investigation revealed that the c.84C>A mutation found in the first allele of the index case was inherited from the father and that the g.347_372del26 deletion in the second allele was a de novo mutation.

Discussion

APRT is a purine salvage enzyme present in all human cells that is encoded by a gene located on chromosome 16q24 (Tischfield and Ruddle 1974; Simmonds 2003), which was cloned in 1987. It normally catalyzes the conversion of adenine to adenosine monophosphate but, in the case of APRT deficiency, adenine is catabolised to 2,8-DHA, a highly insoluble substance that is responsible for crystalluria, nephrolithiasis, and intratubular deposition with or without symptoms. The subsequent renal damage is characterized by chronic tubulointerstitial nephritis with progressive sclerosis and chronic renal failure until the terminal phase, which usually develops during adulthood. The treatment of this disease consists of hydropinic therapy and a low purine diet combined with a daily 10 mg/kg dose of allopurinol. Unlike in the case of other renal stones, it is not necessary to acidify or alkalinize urine because 2,8-DHA precipitates at every pH. The therapy is relatively simple and effective and, in many cases, its prompt administration stops damage progression (Edvardsson et al. 2001; Bollée et al. 2010; Stratta et al. 2010).

This seems to be our patient's case because the micropapillary deposits disappeared in less than 6 months and no new stones have formed but, in many other published cases, the disease was not diagnosed until the kidney had been irreversibly damaged (Bollée et al. 2010; Edvardsson et al. 2001) or even after a recurrence in a transplanted kidney (Benedetto et al. 2001; Bollée et al. 2010; Edvardsson et al. 2001; Stratta et al. 2010). The diagnostic phase is therefore crucial, and the extremely small number of cases reported by now is in contrast with a homozygosity at APRT locus estimated between 1 in 50,000 and 1 in 100,000. The most plausible explanation for this was that APRT deficiency may be largely unrecognized by clinicians involved in the treatment of patients with urolithiasis and/or renal failure (Bollée et al. 2010; Bouzidi et al. 2007; Ceballos-Picot et al. 1992) .

We believe that APRT deficiency should be suspected in every child with renal stones, especially if they are radiolucent and there is no evidence of hyperuricuria, as in our case. The stones may occasionally be radiopaque because of the presence of calcium salts, but the diagnosis is suggested by examining the urinary sediment, which shows 2,8-DHA crystals (Bouzidi et al. 2007; Ceballos-Picot et al. 1992) in about 96% of untreated patients (Edvardsson et al. 2001). These crystals are typically round and reddish-brown with a Maltese cross pattern under polarized light, but an experienced laboratory is required to recognize them (Bouzidi et al. 2007).

In our case, the diagnosis was directly suggested by the spectrophotometric stone analysis. The other methods usually used in clinical practice, such as a murexide test, colorimetry reaction or thermo-gravimetric analysis cannot distinguish them from uric acid stones (Ceballos-Picot et al. 1992).

The diagnosis can be confirmed by measuring APRT activity in red blood cells and undertaking a genetic analysis. In our rare case, the disease manifested itself in very early infancy, but we cannot say whether the de novo mutation may have been responsible for its early clinical expression. The familial genetic study showed that the father and sister were heterozygous carriers of the same

previously reported mutation and, respectively, showed 7% and 38% residual enzymatic activity. It is interesting to note that despite this low level (especially in the father) there was no evidence of crystalluria or any US signs of nephrocalcinosis or stones, thus confirming that heterozygotes are usually completely asymptomatic.

In conclusion, pediatric lithiasis (especially in the first years of life) is a relatively rare condition, and may be a major indicator of high morbidity congenital or metabolic disorders that could be diagnosed and treated, sometimes with good clinical results. For this reason, a multidisciplinary approach is always mandatory in every case of child with urinary stones: obviously, the first step in the management of these patients is to relieve pain and undertake appropriate imaging to enable a decision as to whether and how to remove the stones or resolve obstruction. The second step, similarly mandatory, is the metabolic evaluation. APRT deficiency can be diagnosed by means of a urinary sediment examination, an affordable and reliable method that should represent the first diagnostic step in every child with lithiasis. However, a correct diagnosis requires a high index of suspicion that is possible only with a good knowledge of the disease.

Support: the molecular studies for the diagnosis of APRT and HPRT deficiencies are supported by the Association Lesch-Nyhan Action and the Association Malaury.

Dr Vercelloni's work has been sponsored by a research fund "Ricerca Corrente anno 2011" thanks to Fondazione IRCCS Ca'Granda Ospedale Maggiore Policlinico Milano.

Acknowledgements We thank Véronique Droin for her technical assistance in APRT enzymatic activity determination.

Take-Home Message

Pediatric lithiasis (especially in the first years of life) is a relatively rare condition, and may be a major indicator of high morbidity congenital or metabolic disorders that could be diagnosed and treated, sometimes with good clinical results. For this reason, a multidisciplinary approach including a scrupulous metabolic evaluation is always mandatory in every case of child with urinary stones. APRT deficiency can be diagnosed by means of a urinary sediment examination, an affordable and reliable method that should represent the first diagnostic step in every child with lithiasis. However, a correct diagnosis requires a high index of suspicion that is possible only with a good knowledge of the disease.

References

Benedetto B, Madden R, Kurbanov A, Braden G, Freeman J, Lipkowitz G (2001) Adenine phosphoribosyltransferase deficiency and renal allograft dysfunction. Am J Kidney Dis 37(5): E37

Bollée G, Dollinger C, Boutaud L, Guillemot D, Bensman A, Harambat J, Deteix P, Daudon M, Knebelmann B, Ceballos-Picot I (2010) Phenotype and genotype characterization of adenine phosphoribosyltransferase deficiency. J Am Soc Nephrol 21(4):679–688

Bouzidi H, Lacour B, Daudon M (2007) 2,8-Dihydroxyadenine nephrolithiasis: from diagnosis to therapy. Ann Biol Clin (Paris) 65(6):585–592

Ceballos-Picot I, Perignon JL, Hamet M, Daudon M, Kamoun P (1992) 2,8-Dihydroxyadenine urolithiasis, an underdiagnosed disease. Lancet 339:1050–1051

Edvardsson V, Palsson R, Olafsson I, Hijaltadottir G, Laxdal T (2001) Clinical features and genotype of adenine phosphoribosyltransferase deficiency in Iceland. Am J Kidney Dis 38(3):473–480

Simmonds HA (2003) Adenine phosphoribosyltransferase deficiency. Orphanet Encyclopedia. www.orpha.net

Stratta P, Fogazzi GB, Canavese C, Airoldi A, Fenoglio R, Bozzola C, Ceballos-Picot I, Bollée G, Daudon M (2010) Decreased kidney function and crystal deposition in the tubules after kidney transplant. Am J Kidney Dis 56(3):585–590, Epub 2010 Mar 19

Tischfield JA, Ruddle FH (1974) Assignment of the gene for adenine phosphoribosyltransferase to human chromosome 16 by mouse-human somatic cell hybridization. Proc Natl Acad Sci U S A 71:45–49

JIMD Reports
DOI 10.1007/8904_2011_93

INVITED ARTICLES

Alkaptonuria: Leading to the Treasure in Exceptions

Timothy M. Cox

Received: 1 June 2011 /Revised: 28 August 2011 /Accepted: 8 September 2011 /Published online: 6 December 2011
© SSIEM and Springer-Verlag Berlin Heidelberg 2011

Abstract The brilliant geneticist, William Bateson, a formidable English experimentalist, was the first to recognize the nature of the "inborn" in Archibald Garrod's errors of metabolism. Bateson's advice to young scientists: "Treasure your exceptions!" summarizes much of the vigorous empiricism associated with the study of rare disorders.

The first inborn error of metabolism to be so recognized was alkaptonuria, and it is only recently that a proper understanding of this condition as a disease, rather than a biochemical curiosity, has emerged. Abnormal excretion of the reactive tyrosine metabolite, homogentisic acid, not only provides a tangible biomarker of alkaptonuria, but also a focus for detailed mechanistic understanding.

Currently, there is no proven treatment for alkaptonuria but emergence of orphan drug legislation internationally has promoted the licensing of nitisinone (Orfadin™) for an equally rare disorder of tyrosine metabolism – hereditary tyrosinaemia type 1. Nitisinone, a triketone competitive inhibitor of a proximal step leading to the formation of homogentisic acid, has potent therapeutic effects in hereditary tyrosinemia and rapidly ameliorates the primary biochemical abnormality in patients with alkaptonuria.

Here, we discuss the context in which nitisinone should be further explored for the treatment of alkaptonuria. This exceptional disease is a paradigm case, which opens up unusual opportunities for basic and applied research. In modern times, it also shows how the conflation of orphan drug legislation and the emerging power and commitment of patient organizations can synergize effectively to advance basic research and therapeutic development in ultra-orphan diseases.

William Bateson and His Influence: Treasure Your Exceptions

Archibald Garrod (1857–1936), whose clinical research into alkaptonuria led to the definition of "Inborn Errors of Metabolism" (Garrod 1901, 1902, 1909) is the icon of those who hold the study of inherited metabolic diseases dear. But Garrod – and posterity – owe much scientifically and philosophically to his near contemporary William Bateson (1861–1926).

In the captivating era of molecular genetics, Bateson is now largely forgotten but not only he was the amanuensis, champion and biographer of Mendel in anglophone science (Harper 2005), he invented many useful and everyday terms, including heterozygote, homozygote, and allelomorph.

William Bateson and colleagues were the first to identify the phenomenon of genetic linkage and teased out the influence of what were termed epistatic factors (Saunders et al. 1906, 1908; Bateson 1913), as well as homeosis – the latter, describing determinants controlling segmentation and patterning in development. Bateson referred to the process by which one body part is transformed into the likeness of another, from the Greek, homoios = same and -osis = condition, or process (Bateson 1894, 1913). Mutations in homeotic genes are now recognized as unitary determinants underlying many complex developmental disorders in plants and animals – and with emerging significance in humans (Boncinelli 1997). It was indeed William Bateson who indicated to Garrod that the pattern of inheritance and

Communicated by: Jean-Marie Saudubray.

Competing interests: None declared.

T.M. Cox (✉)
University of Cambridge, Cambridge, UK
e-mail: tmc12@medschl.cam.ac.uk

frequent consanguinity in pedigrees affected by alkaptonuria immediately suggested transmission of a unitary Mendelian recessive character – thus explaining the nature of the *inborn* in "errors of metabolism" (Saunders and Bateson 1902; Garrod 1931).

In his essay, "The Debt of Science to Medicine" (Garrod 1924) based on his Harveian Oration, Archibald Garrod refers to a much-quoted statement by William Harvey (1578–1657) in correspondence sent by Harvey within a few weeks of his own death to a Dutch physician:

> It is even so – Nature is nowhere more accustomed more openly to display her secret mysteries than in cases where she shows traces of her workings apart from the beaten path; nor is there any better way to advance the proper practice of medicine than to give our minds to the discovery of the usual law of Nature, by careful investigation of rarer forms of disease For it has been found, in almost all things, that what they contain of useful or applicable is hardly perceived unless we are deprived of them, or they become deranged in some way
> (Harvey, trans. Wills 1847).

Garrod was in several senses Harvey's latter day apostle: a clinical experimentalist in the study of inborn errors of metabolism, he was the advocate of a powerful investigative tradition within medicine (Garrod 1928). But Harvey's principle, at least in English, has more pithy echoes elsewhere, including those in the contemporaneous writings of Francis (later Lord) Bacon:

> For he that knows the way of nature will more easily observe her deviations: and on the other hand, he that knows her deviations will more accurately describe her ways
> (Bacon 1620).

However, the sentiment is probably best crystallized by William Bateson:

> *Treasure Your Exceptions!*
> (Bateson 1908).

Bateson's research had led to the identification of genetic linkage ("coupling") in the Sweet pea, *Lathyrus odoratus* – a phenomenon that by definition had escaped Mendel's attention because the seven characters he studied in pure-bred strains of the Garden pea, *Pisum sativum,* in practical terms segregated independently. William Bateson (working with Saunders and Punnett) explored the apparently anomalous transmission of several characters, including flower pigments, in the Sweet pea; he also identified and characterized the phenomenon of epistasis (Saunders et al. 1905).

Epistasis and the Biochemical Genetics of Metabolic Processes

George Beadle and Edward Tatum, who studied radiation-induced auxotrophic mutants in the red bread mould, *Neurospora crassa*, are credited with the "one gene – one enzyme" principle, and were awarded the Nobel prize in 1958. Initial studies by Beadle with Boris Ephrussi using experimental genetics specifically to interrogate biochemical pathways, examined genetic determinants of eye pigments in *Drosophila.* Ephrussi, had perfected the use of transplantation to transfer embryonal eye tissue from one larva to another. By studying the eye after maturation in the adult *Drosophila* derived from tissue originating in larvae of defined strains with mutant eye-colours, Ephrusssi and Beadle embarked on a study of ocular pigmentation in the fly.

In nearly all cases of the 26 separate eye-colour mutants of *Drosophila* studied, the colour of the transplanted eye was determined by the stock from which it was derived; but in two instances, *vermilion* and *cinnabar* (determined by recessive genes), the eye discs transplanted from the mutant larvae into wild-type recipients did not assume the distinctive red colour of the donor – rather the eyes were wild type (brown). Ephrussi and Beadle deduced that a diffusible substance in the wild-type recipient strain could complement the defective formation of pigment in the transplanted mutant *vermillion* and *cinnabar* tissue. To determine the putative precursor–product relationships between these two mutant strains, embryonal eye tissue was exchanged between them: painstaking experiments categorically showed that while there was complementation of *vermillion* by *cinnabar*, the red cinnabar pigment, due to incomplete formation of brown pigment was not corrected in transplants from *cinnabar* into *vermillion* larvae.

These experiments showed that the eye-colour genes determined distinct metabolic steps in the cognate biochemical processes leading to eye pigment formation. It took several years to identify the metabolic disturbances and biosynthetic steps in the formation of the eye pigments: Ephrussi eventually found that feeding tryptophan overcame the defect in *vermillion* but it took several years and intensive biochemical study by others, including Butenandt and Kuhn, to work out the chemical reactions leading to the formation of kyurenine (the product of the wild-type vermillion gene) and its derivatives such as 3-hydroxykyurenine (the product of the wild type *cinnabar* gene) as immediate precursors to the complex brown eye pigment (Beadle 1958).

Although theoretically it is possible to test the effect of any genetic combination on a given biochemical process, the work of Beadle and Ephrussi illustrates the experimental

 Springer

difficulties that arise in dissecting complex metabolic reactions since a fly that was homozygous for the vermillion could also harbour two copies of the mutant cinnabar gene but be indistinguishable from vermillion since the effect of the cinnabar mutation occurred after the block in pigment biosynthesis (Beadle and Ephrussi 1937). This latter phenomenon, known from the work of Bateson as epistasis, has since been exploited productively to investigate countless metabolic pathways.

Characterizing the precursors of the eye pigment in *Drosophila* was laborious and held too many uncertainties for further productive study of the chemical disturbances underlying genetic abnormalities of metabolism. While acknowledging the insight shown by Garrod in his happy choice of alkaptonuria (in which the chemistry of homogentisic acid had been long established), Beadle and Tatum inverted the logic of their thesis that enzyme-catalysed reactions were specified by individual genes: they selected a model organism with simple nutritional requirements, the Ascomycete mould, *Neurospora. Neurospora,* a haploid organism which can reproduce sexually, can easily be manipulated for genetic study and is susceptible to chemical and radiation mutagenesis. Moreover, since the organism is prototrophic (able to generate complex biomolecules from simple inorganic precursors), the growth requirements of its auxotrophic mutants are readily characterized by supplementing minimal growth media with defined constituents – thus facilitating identification of genetic defects in essential metabolic pathways. The four products of a single meiotic process in Neurospora can be individually recovered, and this allows the investigator to determine whether the induced nutritional deficiencies are caused by mutations in single genes. If they are, crosses with the original should yield four mutant and four non-mutant spores of the eight products in each spore sac.

In a test case, Beadle and Tatum found that spores from several mutant strains would not replicate without addition of arginine to the culture (arginine-*less*). Distinct arginine-*less* mutants were isolated, each lacking the function of a specific gene; since compounds such as citrulline and ornithine were known metabolites of arginine in higher organisms, their ability to substitute for arginine was tested. Only specific combinations restored growth in each arginine-*less* strain, and thus the metabolic block in the individual mutants could be identified allowing the individual steps in arginine biosynthesis to be defined. In this way, the one gene – one enzyme concept was promulgated. The experimental approach has allowed innumerable metabolic pathways to be characterized, and remains at the heart of classical biochemical genetics.

In his Nobel lecture, Beadle acknowledged the work of Archibald Garrod (Beadle 1958) but in reality the mechanism involving the operation of epistatic factors was discovered and annunciated in detail by William Bateson. By persistently exploring exceptions in the inheritance of flower colour in the Sweet pea and other species, Bateson and E. Rebecca Saunders had teased out the phenomenon (Saunders et al. 1906, 1908). Crossing two white varieties of Sweet pea unexpectedly produced purple-flowered offspring; and when these first-generation plants were self-fertilized, various second-generation offspring, with flowers ranging from red to purple were obtained – but the progeny also included plants with white flowers.

The phenomenon of epistasis was revealed by analysing these apparently anomalous patterns of inheritance: Bateson deduced not only that homozygosity for two distinct genes was responsible for white flowers, but that inheritance of one or more dominant alleles in the hybrids, which were genetic compounds, led to the formation of different pigments mediated by the action of discrete enzymes ("ferments") on precursor substances (Bateson 1913). These proved to be a series of metabolites involved in the biosynthesis of the purple floral pigment (Wheldale 1915). Later, with the interested support of Bateson, and JBS Haldane in Cambridge, Muriel Onslow (née Wheldale) conducted breeding experiments to dissect the unitary steps of anthrocyanin pigment formation in the snapdragon and other plants (Wheldale Onslow et al. 1981).

While Beadle acknowledged that the line of inquiry into pigment biosynthesis pursued by Wheldale was important and that: "It began with a genetic study of flower pigmentation in snapdragons. But soon the genetic observations began to be correlated with the chemistry of the anthocyanin and related pigments that were responsible". The material was favourable for genetic and chemical studies and the work has continued almost without interruption to yield new information ever since. It is surprising that as a geneticist Wheldale did not acknowledge the primary influence and inspiration of Bateson.

Apart from alkaptonuria, the original inborn errors described in humans by Garrrod were the pigment disorder albinism, cystinuria and pentosuria – later followed by congenital porphyria. Asbjørn Følling's masterful chemical analysis was critical to the discovery of phenylpyruvic imbecility (phenylketonuria), and a spectacular affirmation of the experimental approach to biochemistry adopted by Garrod (Følling 1934). Coincidentally, albinism, alkaptonuria and phenylketonuria turn out to be genetically distinct defects in the metabolism of human aromatic amino acids. The occurrence of red or blond hair with progressive skin hypopigmentation (albinism) and blue eyes in patients with phenylketonuria is a curious illustration of epistatic phenomena in human aromatic amino acid metabolism, since the action of tyrosinase is impaired. Type 1 hereditary tyrosinaemia is also a recessive disorder affecting this pathway; as

described below, its relationship to alkaptonuria is therapeutically informative.

The principles of biochemical genetics, axiomatic in the contemporary study of metabolic diseases, emerged from genetic studies of epistasis described by William Bateson; but the treasure in his exceptions extends far beyond the scientific laboratory and textbook. The definitive extension of Bateson's research outside classical genetics showed that all biochemical processes in all organisms are under genetic control, that these overall biochemical processes are resolvable into a series of individual stepwise reactions, and that single reactions are controlled in a primary fashion by a single gene (Tatum 1958). Biochemical mutants have led the study of innumerable complex biosynthetic and catabolic pathways, but the detailed characterization of inborn errors and hereditary developmental abnormalities in humans has informed the spheres of biological science through medicine in a manner predicted by Garrod but now on an unprecedented scale.

Alkaptonuria: A Disease Rather than a Biochemical Curiosity

Garrod convincingly demonstrated that alkaptonuria, the first inborn error of metabolism to be recognized, is due to an hereditary defect in the scission of the benzene ring at an intermediate step in the breakdown of the amino acid, tyrosine (Garrod 1901, 1902, 1909). It took 50 years before Garrod's hypothesis was verified by the resolution of aromatic amino acid degradation into six enzymatic steps occurring via homogentisic acid (2,5-dihydroxyphenylacetic acid). We now know that the oxidation of homogentisic acid by the iron-containing homogentisic acid oxidase, EC 1.13.11.5 (which yields maleylacetoacetic acid in the fourth reaction of the degradative pathway) is defective in alkaptonuria (Fernandez-Canon et al. 1996; Mitchell et al. 2001; Chakrapani and Holme 2006). The characteristic feature of the disorder, which in most countries has an estimated birth frequency of 1 in 200,000–1,000,000, is excess urinary homogentisic acid; when oxidized, this molecule generates the pathognomic black polymerized product. This so-called ochronotic pigment appears in several organs, including the prostate gland; it also accumulates preferentially in collagen-rich tissues, including cartilage.

In the first rigorous study of the natural course of alkaptonuria, Phornphutkul et al. (2002) showed that, far from representing a curious biochemical abnormality as accepted in Garrod's day, the condition shortens life as a result of premature heart valve calcification and coronary heart disease; moreover, renal calculi are frequent and a painful degenerative arthropathy of major joints, including those of the spine, greatly impair mobility and life quality (Mitchell et al. 2001/2; Mannoni et al. 2004; Perry et al. 2006; Chakrapani and Holme 2006).

Although the molecular pathogenesis of these complications is not fully understood, characteristically the damaged tissues are heavily pigmented, indicating a relationship between the pigment derived directly from homogentisic acid and the cellular injury (Mannoni et al. 2004). Mammalian cartilage contains polyphenyl oxidases, which can catalyse the oxidation of homogentisic acid into pigment, and while benzoquinone acetic acid has been identified as a potential oxidative intermediate in alkaptonuria, deposits are found within ochronotic chondrocytes. It remains to be determined whether the pigment deposition and binding to extracellular matrix occurs primarily at the intracellular or at the extracellular level.

Treatment: Alleviating the Effects of Alkaptonuria

There is no proven treatment for alkaptonuria (Mitchell et al. 2001/2; Mannoni et al. 2004; Chakrapani and Holme 2006). Dietary modifications to restrict the ingestion proteins rich in tyrosine and phenylalanine have met with limited biochemical success and pose difficulties in the long-term owing to the need to ensure lifelong strict control. Interfering with the ochronotic process by, for example, supplying a reducing agent such as ascorbic acid has been proposed – but the enormous overproduction of homogentisic acid, the precursor of the harmful metabolite, suggests that only a large decrease in the pathological supply of toxic molecules would afford benefit.

Sharing Amongst Orphans in the Tyrosine Degradation Disorders

Studies in another inborn error of metabolism, also affecting the pathway of tyrosine degradation, are illuminating. Hereditary tyrosinemia type 1 is an extremely rare autosomal recessive disease caused by deficiency of the distal enzyme fumarylacetoacetate hydrolase (EC 3.7.1.2) (Mitchell et al. 2001/2). This defect leads to a failure of the final step in the breakdown of tyrosine and fumaryl acetoacetate accumulates in parenchymal liver cells and the epithelium of the proximal renal tubule. Fumaryl acetoacetate is a carcinogen and induces alkylation of DNA: liver enlargement with growth failure and progressive disturbances of kidney and liver function occur with cirrhosis and a high risk of hepatocellular carcinoma.

 Springer

Fanconi syndrome with renal tubular acidosis may develop; and with the accumulation of the secondary metabolite, succinylacetone (4,6: dioxoheptanoate, a potent inhibitor of 5-aminolaevulinate dehydratase), neurovisceral episodes mimicking acute porphyria occur (Mitchell et al. 1990, 2001/2; Sassa and Kappas 1983).

With the introduction of the Orphan Drug legislation in the United States in 1983, and in Europe in 2000, the horizon is fast changing (Haffner 2006; Braun et al. 2010; European Committee for Orphan Medicinal Products 2011). The European legislation in relation to orphan medicinal products has been of direct value to the small community of patients affected by hereditary tyrosinaemia. One of the early successes of orphan drug legislation in Europe was the designation of nitisinone[1] for hereditary tyrosinemia type 1 granted to Swedish Orphan. Nitisinone ("Orfadin™") is an early orphan drug developed from a triketone herbicide preparation which was found serendipitously to be a reversible competitive inhibitor of 4-hydroxyphenylpyruvate oxidase, the second enzyme in the breakdown of tyrosine (third enzyme in the disposal of phenylalanine) – thus decreasing the distal formation of malelyacetoacetic acid after the benzene ring has been opened by the action of homogentisic acid oxidase. Malelyacetoacetic acid is isomerized to fumarylacetoacetic acid, the principal toxic metabolite in type 1 tyrosinaemia (Lindstedt et al. 1992; Lock et al. 1998). Nitisinone is generally well tolerated by mouth; rare unwanted effects include leukopenia and thrombocytopenia with increased plasma tyrosine concentrations, which may cause skin irritation and reversible lenticular changes; the agent is thus combined with a tyrosine- and phenylalanine-restricted diet (McKiernan 2006).

Nitisinone was designated promptly as an orphan medicinal product in 2000 for the treatment of tyrosinaemia type 1. Currently supplied by Swedish Orphan Biovitrum AB, it received marketing approval in February 2005. Nitisinone transforms the course of hereditary tyrosinaemia, improving patients with established liver disease and reversing the non-hepatic manifestations: in infants and adults, the drug is used at a dose of 0.5–2 mg/kg body weight in divided doses daily. Introduced in infancy, nitisinone provides protection against development of hepatocellular carcinoma (Chakrapani and Holme 2006; Masurel-Paulet et al. 2008; Santra and Baumann 2008) but the risk of this fatal cancer remains if the drug is not started before the age of 2 years (McKiernan 2006).

First Clinical Trials in Alkaptonuria

With a specific competitive inhibitory action at the level of 4-hydroxyphenylpyruvate oxidase an early step in the degradation of tyrosine, nitisinone has an obvious application in alkaptonuria (Anikster et al. 1998; Masurel-Paulet et al. 2008). Clearly, inhibiting the enzyme that catalyses the step preceding homogentisic acid oxidase would decrease accumulation of the metabolite that serves as a precursor of the ochronotic pigment (Anikster et al. 1998; Suzuki et al. 1999), 4-hydroxyphenylpyruvate oxidase (EC 1.2.3.13) is a primary therapeutic target in alkaptonuria. Studies by William Gahl and colleagues at the National Institutes of Health showed the rapid effect of low-dose nitisinone on homogentisic acidaemia and homogentisic aciduria in alkaptonuric subjects (Phornphutkul et al. 2002; Suwannarat et al. 2005). Moreover, the dose required to alleviate the biochemical abnormalities appears to be approximately $\sim$2% of that required for hereditary tyrosinemia Type 1, disease due to mutations in the last step of the pathway of tyrosine breakdown.

In 2003, Swedish Orphan received a positive opinion for orphan designation of nitisinone for the treatment of alkaptonuria from the Committee for Orphan Medicinal Products. With this background, a long-term study of nitisinone to treat alkaptonuria was undertaken at the National Institutes of Health Clinical Centre (NCT00107783) sponsored by the National Human Genome Research Institute (NHGRI). The 3-year phase I/II trial to examine the safety and effectiveness of long-term nitisinone for joint disease enrolled 40 alkaptonuric patients, of whom 20 were randomized to placebo in addition to their regular medication, and 20 took regular medication plus 2 mg of nitisinone daily. Serial measurements of liver-related tests, clinical examinations, 24-h determinations of urine metabolites, blood and urine tyrosine and determination of other amino acids, homogentisic acid, skeletal radiology, spiral computerized X-ray tomography of the abdomen to detect kidney stones, and ophthalmoscopy.

Completed in April 2009, the primary outcome measure of hip joint range of motion was not met at the end of the trial. Nonetheless, nitisinone was well tolerated and greatly reduced urinary homogentisic acid excretion in the urine from 4 ± 1.8 to 0.2 g/day (normal $< 0.028 \pm 0.015$ g/day). As expected, plasma tyrosine concentrations rose approximately from 70 to 760 ± 181 μM with no ill-effects attributable. At dose administered and as predicted, nitisinone was well tolerated; and although the trial failed its primary therapeutic endpoint, improvement of joint symptoms in some patients with established joint disease was identified (Introne et al. 2011).

---

[1] Nitisinone: 2-(2-nitro-4-trifluoromethylbenzoyl) cyclohexane-3-dione-2-2-nitro-4-trifluoromethylbenzoyl cyclohexane-1-3-dione.

Alkaptonuria and the Complexities of Orphan Therapeutics

The design of clinical trials for rare diseases with few participants, and uncertainties about what outcomes will be informative over long periods of observation, is very challenging (Haffner 2006; Griggs et al. 2009; European Committee for Orphan Medicinal Products 2011). Alkaptonuria, a neglected ultraorphan disease and the first-identified inborn error of metabolism, represents an extreme case for therapeutic advance. While in retrospect, it is not surprising that nitisinone failed to increase hip mobility in alkaptonuric patients with established ochronotic arthropathy, given the definitive effects of nitisinone on the primary biochemical defect in alkaptonuria, it is difficult to ignore the potential for clinical utility (Phornphutkul et al. 2002; Suwannarat et al. 2005; Introne et al. 2011). However, the problem of how best to investigate its putative therapeutic and more likely preventative action on disease-related events that are meaningful for patients – and obtain appropriate funding for long-term studies and regulatory approval – remains (Buckley 2008; Griggs et al. 2009).

The incentives provided by the orphan drug legislation clearly provide the background that will be critical for development. In the first 25 years since introduction of the United States Orphan Drug Act, about 250 drugs were approved for more than 200 rare diseases. In the 20 years since the inception of the regulation on orphan medicinal products in Europe, there were more than 1,200 applications and more than 850 of these have been designated as orphan medicinal products – it is encouraging to see the number of applications continues to rise (Braun et al. 2010; European Committee for Orphan Medicinal Products 2011). Taking advantage of an accelerated time between designation as an orphan medicinal product and marketing authorization, the mean lapse of time is less than 3 years.

Despite all this activity however, the number of products receiving marketing authorization is very much smaller. Many of the new products include those for somatic cell therapy, tissue engineered medicinal products and gene therapy – only about one-third of those designated can be regarded as "innovative," that is utilizing emerging techniques and molecular technology (including monoclonal antibodies and soluble receptor fragments). In contrast, nitisinone is a small molecule relatively cheap to manufacture, with defined biochemical effects and not, at first sight, in the innovative class (European Committee for Orphan Medicinal Products 2011). In the commercial world, exclusive marketing is a key incentive, but the intention of the regulation by the EC is clearly to obtain more and better drugs for rare diseases, rather than to create

monopolies (European Committee for Orphan Medicinal Products 2011). In fact, marketing exclusivity can be challenged in cases of lack of supply, proven clinical superiority – or agreements to share the market with an original sponsor. It is notable that one of the early marketing approvals – enzyme replacement therapy for Fabry disease – was granted by the European Medicines Agency to two recombinant enzyme products made by two competing companies on the same day!

Further advantages of applications for designation of orphan medicinal products include "protocol assistance" from the Committee for Orphan Medicinal Products: this benefit is increasingly attractive for companies seeking designation – indeed its acceptance appears to be highly correlated with successful marketing authorization. These features, marketing authorization with global revenue and the prior designation of nitisinone as an orphan medicinal product for investigation in alkaptonuria, seem to provide a strong basis for continued investment.

The Emergence of Therapeutic Consortia Driven by Patients

A striking parallel development in the evolution of medicinal products for rare diseases has been the emerging power and commitment of patient organizations. This corollary of the commercial investment from pharmaceutical and Biotech industries also signals radical changes in the relationship between clinicians, professional researchers and patients. Even the longest established patient organizations representing the interests of patients and their families affected by rare diseases were founded in the middle or latter half of the twentieth century at the earliest, but at the same time their direct influence on therapeutic developments has grown sharply. The occurrence of but a few patients and their families affected by rare diseases can initiate workable collaborations between professional researchers, including clinicians, for the advancement of particular scientific agendas. Not only can these collaborations drive strong research agendas, but they can form powerful lobbying groups for negotiation with funding agencies and industry (Ingelfinger and Drazen 2011). A further aspect is that while the individual disease may be rare, clinicians and other professionals with expertise and interest are usually even less frequent: thus patient-led collaborations combine a unity of purpose with invaluable economy of scale.

There have been many instances where new treatments have been explored through this pathway: perhaps, the most renowned was the success of Augusto and Michaela Odone in the development of Lorenzo's oil for patients with adrenoleukodystrophy (Moser et al. 1987). Other examples

 Springer

where patient groups have sponsored clinical and basic research and facilitated access to rare patients who take part in clinical trials include Gaucher disease and cysteamine in cystinosis - as well as the recent case of the use of sirolimus (rapamycin) in lymphangioleiomyomatosis (Ingelfinger and Drazen 2011).

Alkaptonuria: "FindAKUre"

Patients and their families form associations dedicated to research into treatments; they have often been successful in extending the licence for other conditions. In many instances, the organizations have supported basic research and seen to it that this is regulated by high standards of peer review. Patient organizations and their related charities are able to bring in clinicians and laboratory investigators and broker productive collaborations. In many ways, the former paternalistic traditions of medicine tend to be outmoded; and now doctors are seen more as colleagues with useful skills, which can be harnessed in a common effort to improve the outcomes of challenging diseases.

With the unmet needs of patients with alkaptonuria, a Europe-wide, multi-disciplinary scientific collaboration has been assembled with independent charitable funding to find an effective treatment for this disease (http://www. findakure.org/contact_us.php). Given that the frequency of alkaptonuria in countries such as the Dominican Republic and Slovakia is much greater than in most populations (in Slovakia, about 1 in 19,000, Srsen et al. 2002), the international links will be mutually attractive for investigation.

Originally based at the University of Liverpool, UK and a venture initiated jointly with the UK Alkaptonuria Society and the University of Liverpool, a key element of the "FindAKUre" project is the engagement of leading exponents in clinical, animal, laboratory, and bioinformatics research in a strategy that is focussed on the pathological and molecular understanding of ochronosis at all levels from extended natural history studies to translational therapeutics. The international nature of this collaboration, already including experts from Slovakia, has the capacity to provide access of several hundred patients and thereby greatly expand the scope of investigations – and may prove critical for decisions in relation to funding for new therapeutic trials.

Treasure Your Exceptions: Envoi

Archibald Garrod's "Debt of Science to Medicine" (Garrod 1924) still holds, and it is only now that signs of repayment are visible. The arduous search for so-called "disease genes" has always been predicated on the promise of therapeutic innovation for each of the thousands of inherited diseases (Dietz 2010). In the first 100 years after Garrod's discovery of the inborn errors and despite comprehensive studies of their biochemical genetics, beyond nutritional manipulation and co-factor supplementation, few therapeutic innovations were brought into clinical practice. Now understood as an authentic disease, and emblematic of countless inborn errors of metabolism, alkaptonuria is a very rare (ultra-orphan) condition and despite all that we know of its biochemistry and molecular genetics, it persists frustratingly at the threshold of innovative therapeutics.

The recent crisis in drug discovery experienced by practically all major pharmaceutical companies paradoxically gives ground for hope that more courageous investment in informative rare diseases will be forthcoming. There has been a spate of acquisitions by "Big Pharma" of smaller but successful biotechnology companies (Sharma et al. 2010; Dolgin 2010): biotechnology companies may be rich with scientific ideas and avenues for discovery, but without the expertise and resources fully to develop these qualities and yield useful drugs, they remain an empty platform for experimental and translational medicine in the field of rare disease – "orphans unadopted".

Comprehensive treatment of alkaptonuria and type I tyrosinaemia, including the clinical use of nitisinone, is emerging through a thorough understanding of tyrosine metabolism. This owes its origin to the classical studies of Garrod and Bateson. It is now clear that successful innovation in treating alkaptonuria will depend on the emerging power of collaborations between patient-based organizations, professional groups and industry. Formidable regulatory (and hence funding) obstacles lie ahead and strong advocates will be needed to ensure that even the favoured agent, nitisinone, can be evaluated wisely for prophylaxis in the face of an existing failed phase II trial. Long-term studies after early introduction of the drug will be critical to success but will present conceptual challenges for regulators who are unwilling to accept evidence from any other source than the randomized double-blind cross-over clinical trial. For ultra-orphan diseases, and where blinded control studies are impossible (for example, urinary changes preclude this option in alkaptonuria) unconventional trial designs and long-term observational studies will be of a crucial value.

For alkaptonuria, a rich vein of treasure is yet to be found within the old mine of classical discovery in genetics and biochemistry. Rewarding exploration of this treasure will depend as much on the need to succeed as it will on the contemporary availability of powerful molecular tools, translational scientists and clinicians for recruitment to the field.

The combination of strong commercial incentives with better–informed advocacy on behalf of patients gives hope that the full potential of orphan drug legislation will eventually be realized. Taking the example of the European Organization for Rare Diseases – a non-governmental patient-driven alliance of patient organizations and individuals – this organ, has with others been instrumental in driving the adoption of the European Council recommendation (2009/C 151/02) for comprehensive National action in the field of rare diseases in each constituent country of Europe by 2013. Although these are laudable initiatives, it should be realized that the critical needs of countless patients worldwide with innumerable rare diseases remain unmet. More than 95% of rare diseases lack treatments – often when there have been successful therapeutic studies in animal models.

In addition, where potential treatments can be tested, failure to employ the regulations for accelerated approval (which allow the use of surrogate endpoints to achieve marketing approval for a given agent), is frequent. Such failings have recently been emphasized in stark terms (Miyamoto and Kakkis 2011). Brigitta Miyamoto and Emil Kakkis also point out that in many rare diseases there are difficulties in gaining acceptance of novel surrogate endpoints where there has been no therapeutic precedent. Such hurdles are inappropriate and formidable to overcome – especially when combined with the tendency of some personnel in the regulatory agencies, not to accept or to facilitate design of unconventional clinical trials often required for rare diseases.

Alkaptonuria, albeit a rare but well-recognized disease, is a paradigmatic case by which to test the contemporary influence of orphan drug legislation: so far however, the objective of that legislation has failed in this, the first identified inborn error of metabolism. With a complete understanding of the biochemistry of alkaptonuria, and in possession of a powerful, relatively safe and already licenced orphan agent to ameliorate the primary metabolic abnormality, we remain sandwiched between regulators and the pharmaceutical manufacturer – and apparently impotent in the quest to advance therapy. While now there are burgeoning opportunities to develop orphan drugs, to drive marketing and pay for safe and effective treatments, it is clear that strong partnerships and the will to succeed driven by patients and their advocates must be present before victory can be secured. With increasing recognition of their global importance, the bright ray of public feeling refracted through the tiny prism of rare diseases will surely soon be more acutely perceived, and, one hopes, effective.

The original context renders the quotation from William Bateson both sanguine and practical:

If I may throw out a word of counsel to beginners, it is: Treasure your exceptions! When there are none, the work gets so dull that no one cares to carry it further. Keep them always uncovered and in sight. Exceptions are like the rough brickwork of a growing building which tells that there is more to come and shows where the next construction is to be (Bateson 1908).

Conflict of Interest Statement

The author, who serves as Chairman of the Advisory Board of FindAKUre, reports receiving no fees for activities in relation to alkaptonuria or tyrosinaemia and has no financial interest in the licensing of nitisinone or the activities of Swedish Orphan Biovitrum AB. His programme of research into metabolic diseases is supported in part from the National Institute of Health Research through the University of Cambridge, Biomedical Research Centre (Metabolic theme).

References

Anikster Y, Nyhan WL, Gahl WA (1998) NTBC and alkaptonuria. Am J Hum Genet 63:920–921

Bacon F (1620) Francis Bacon: the new Organon, or true directions for the interpretation of nature (Novum Organum). In: Jardine L, Silverthorne M (eds) (2000) Aphorisms, Book II, Chap. 29. Cambridge University Press, Cambridge, pp 102–221

Bateson W (1894) Materials for the study of variation, treated with especial regard to discontinuity in the origin of species. Macmillan, London, pp 85–86

Bateson W (1908) The methods and scope of genetics, Inaugural Lecture delivered 23 Oct 1908. Cambridge University Press, Cambridge

Bateson W (1913) Mendel's principles of heredity, Chaps 4, 5 and 12. Cambridge University Press, Cambridge, pp. 77–87, 88–106, 212–216, 221

Beadle GW (1958) Genes and chemical reactions in Neurospora. In: Nobel Lectures including presentation speeches and laureates' biographies. Physiology or Medicine, 1942–1962. Elsevier, Amsterdam, pp 587–597

Beadle GW, Ephrussi B (1937) Development of eye colors in Drosophila: diffusible substances and their inter-relations. Genetics 22:76–86

Boncinelli E (1997) Homeobox genes and disease. Curr Opin Genet Dev 7:331–337

Braun MM, Farag-El-Massah S, Xu K, Coté TR (2010) Emergence of orphan drugs in the United States: a quantitative assessment of the first 25 years. Nat Rev Drug Discov 9:519–522

Buckley BM (2008) Clinical trials of orphan medicines. Lancet 371:2051–2055

Chakrapani A, Holme E (2006) Disorders of tyrosine metabolism. In: Fernandes S, van den Berghe W (eds) Inborn metabolic diseases, diagnosis and treatment, 4th edn. Springer, Berlin, pp 233–243

Dietz HC (2010) New therapeutic approaches to mendelian disorders. N Engl J Med 363:852–863

Dolgin E (2010) Big pharma moves from 'blockbusters' to 'niche busters'. Nat Med 16:837

European Committee for Orphan Medicinal Products (2011) European regulation on orphan medicinal products: 10 years of experience and future perspectives. Nat Rev Drug Discov 10:341–349

Fernandez-Canon JM, Granadino B, Beltran-Valero de Bernabe D, Renedo M, Fernandez-Ruiz E, Penalva MA, Rodriguez de Cordoba S (1996) The molecular basis of alkaptonuria. Nat Genet 14:19–24. http://www.findakure.org/contact_us.php. Accessed 31 May 2011

Følling A (1934) Über Ausscheidung von Phenylbrenztraubensäure in den Harn als Stoffwechselanomalie in Verbindung mit Imbezillität. Hoppe-Seyler's Zeitschrift Fuer Physiologische Chemie 227:169–176

Garrod AE (1901) About alkaptonuria. Lancet 2:1484–1486

Garrod AE (1902) The incidence of alkaptonuria: a study in chemical individuality. Lancet 2:1616–1620

Garrod AE (1909) Inborn errors of metabolism; the Croonian Lectures delivered before the Royal College of Physicians of London, in June, 1908. Henry Frowde, Hodder and Stoughton, Oxford University Press, London

Garrod AE (1924) The debt of science to medicine, Harveian oration, before the Royal College of Physicians on St Luke's Day 1924. Clarendon, Oxford University, 30p

Garrod A (1928) The lessons of rare maladies: the Annual Oration delivered before the Medical Society of London on May 21st 1928. Lancet 211:1055–1060

Garrod AE (1931) The inborn factors in disease. Clarendon, Oxford University, London, p 65

Griggs RC, Batshaw M, Dunkle M, Gopal-Srivastava R, Kaye E, Krischer J, Nguyen T, Paulus K, Merkel PA (2009) Clinical research for rare disease: opportunities, challenges, and solutions. Mol Genet Metab 96:20–26

Haffner ME (2006) Adopting orphan drugs – two dozen years of treating rare diseases. N Engl J Med 354:445–447

Harper PS (2005) William Bateson, human genetics and medicine. Hum Genet 118:141–151

Ingelfinger JR, Drazen JM (2011) Patient organisations and research on rare diseases. N Engl J Med 364:1670–1671

Introne WJ, Perry MB, Troendle J, Tsilou E, Kayser MA, Suwannarat P, O'Brien KE, Bryant J, Sachdev V, Reynolds JC, Moylan E, Bernardini I, Gahl WA (2011) A 3-year randomized therapeutic trial of nitisinone in alkaptonuria. Mol Genet Metab, E-pub, May, 6

Lindstedt S, Holme E, Lock EA, Hjalmarson O, Strandvik B (1992) Treatment of hereditary tyrosinaemia type 1 by inhibition of 4-hyroxyphenylpyruvate dioxygenase. Lancet 340:813–817

Lock EA, Ellis MK, Gaskin P, Robinson M, Auton TR, Provan WM, Smith LL, Prisbylla MP, Mutter LC, Lee DL (1998) From toxicological problem to therapeutic use: The discovery of the mode of action of 2-(2-nitro-4- trifluoromethylbenzoyl)-1,3-cyclohexanedione (NTBC), its toxicology and development as a drug. J Inherit Metab Dis 42:498–506

Mannoni A, Selvi E, Lorenzini S, Giorgi M, Airo P, Cammelli D, Andreotti L, Marcolongo R, Porfirio B (2004) Alkaptonuria, ochronosis, and ochronotic arthropathy. Semin Arthritis Rheum 33:239–248

Masurel-Paulet A, Poggi-Bach J, Rolland MO, Bernard O, Guffon N, Dobbelaere D, Sarles J, de Baulny HO, Touati G (2008) NTBC treatment in tyrosinaemia type I: long-term outcome in French patients. J Inherit Metab Dis 31:81–87

McKiernan PJ (2006) Nitisinone in the treatment of hereditary tyrosinaemia type 1. Drug 66:743–750

Mitchell G, Larochelle J, Lambert M, Michaud J, Grenier A, Ogier H, Gauthier M, Lacroix J, Vanasse M, Larbrisseau A et al (1990) Neurologic crises in hereditary tyrosinaemia. N Engl J Med 322:432–437

Mitchell GA, Grompe M, Lambert M, Tanguay RM (2001) Hypertyrosinemia. In:Scriver CR, Beaudet AL, Sly WS, Valle D, Vogelstein B (eds) The metabolic and molecular bases of inherited disease (OMMBID), Chap 79, 8th edn. McGraw- Hill, New York, pp 1777–1805. http://www.ommbid.com. Modified 2002

Miyamoto BE, Kakkis ED (2011) The potential investment impact of improved access to accelerated approval on the development of treatments for low prevalence rare diseases. Orphanet J Rare Dis 6:49

Moser AB, Borel J, Odone A, Naidu S, Cornblath D, Sanders DB, Moser HW (1987) A new dietary therapy for adrenoleukodystrophy: biochemical and preliminary clinical results in 36 patients. Ann Neurol 21:240–249

Perry MB, Suwannarat P, Furst GP, Gahl WA, Gerber LH (2006) Musculoskeletal findings and disability in alkaptonuria. J Rheumatol 33:2280–2285

Phornphutkul C, Introne WJ, Perry MB, Bernardini I, Murphey MD, Fitzpatrick DL, Anderson PD, Huizing M, Anikster Y, Gerber LH, Gahl WA (2002) Natural history of alkaptonuria. N Engl J Med 347:2111–2121

Santra S, Baumann U (2008) Experience of nitisinone for the pharmacological treatment of hereditary tyrosinaemia type 1. Expert Opin Pharmacother 9:1229–1236

Sassa S, Kappas A (1983) Hereditary tyrosinemia and the heme biosynthetic pathway. Profound inhibition of delta-aminolevulinic acid dehydratase activity by succinylacetone. J Clin Invest 71:625–634

Saunders ER, Bateson W (1902) The facts of Heredity in the light of Mendel's discovery. Report of the Evolution Committee of the Royal Society 1:135

Saunders ER, Punnett RC, Bateson W (1905) Further experiments on inheritance in sweet peas and stocks. Proc R Soc Lond B 77:236–238

Saunders ER, Punnett RC, Bateson W (1906) Experimental studies in the physiology of heredity. Report to the Evolution Committee of the Royal Society 3:2–11

Saunders ER, Punnett RC, Bateson W (1908) Experimental studies in the physiology of heredity. Report to the Evolution Committee of the Royal Society 4:2–5

Sharma A, Jacob A, Tandon M, Kumar D (2010) Orphan drug: development trends and strategies. J Pharm Bioallied Sci 2:290–299

Srsen S, Müller CR, Fregin A, Srsnova K (2002) Alkaptonuria in Slovakia: thirty-two years of research on phenotype and genotype. Mol Genet Metab 75:353–359

Suwannarat P, O'Brien K, Perry MB, Sebring N, Bernardini I, Kaiser-KupferMI RBI, Tsilou E, Gerber LH, Gahl WA (2005) Use of nitisinone inpatients with alkaptonuria. Metabolism 54:719–728

Suzuki Y, Oda K, Yoshikawa Y, Maeda Y, Suzuki T (1999) A novel therapeutic trial of homogentisic aciduria in a murine model of alkaptonuria. J Hum Genet 44:79–84

Tatum EL (1958) A case history in biological research. In: Nobel Lectures including presentation speeches and laureates' biographies. Physiology or Medicine, 1942–1962. Elsevier, Amsterdam. http://nobelprize.org/nobel_prizes/medicine/laureates/1958/. Accessed 9 July 2011

Wheldale M (1915) The anthocyanin pigments of plants, 1st edn. Cambridge University Press, Cambridge

Wheldale Onslow M et al (1981) The classical period in chemical genetics: recollections of Muriel Wheldale Onslow, Robert and Gertrude Robinson and J. B. S. Haldane Rose Scott-Moncrieff. Notes and Records of the Royal Society of London 36 (1):125–154

Wills R (1847) The works of W. Harvey (Transl). Sydenham Society, London, p 616

JIMD Reports
DOI 10.1007/8904_2011_96

Chaperone-Like Therapy with Tetrahydrobiopterin in Clinical Trials for Phenylketonuria: Is Genotype a Predictor of Response?

Christineh N. Sarkissian* · Alejandra Gamez* · Patrick Scott · Jerome Dauvillier · Alejandro Dorenbaum · Charles R. Scriver · Raymond C. Stevens

Received: 22 December 2010 / Revised: 04 August 2011 / Accepted: 20 September 2011 / Published online: 6 December 2011
© SSIEM and Springer-Verlag Berlin Heidelberg 2011

Abstract Prospectively enrolled phenylketonuria patients ($n = 485$) participated in an international Phase II clinical trial to identify the prevalence of a therapeutic response to daily doses of sapropterin dihydrochloride (sapropterin, KUVAN®). Responsive patients were then enrolled in two subsequent Phase III clinical trials to examine safety, ability to reduce blood Phenylalanine levels, dosage (5–20 mg/kg/day) and response, and bioavailability of sapropterin. We combined phenotypic findings in the Phase II and III clinical trials to classify study-related responsiveness associated with specific alleles and genotypes identified in the patients. We found that 17% of patients showed a response to sapropterin. The patients harbored 245 different genotypes derived from 122 different alleles, among which ten alleles were newly discovered. Only 16.3% of the genotypes clearly conferred a sapropterin-responsive phenotype. Among the different *PAH* alleles, only 5% conferred a responsive phenotype. The responsive alleles were largely but not solely missense mutations known to or likely to cause misfolding of the PAH subunit. However, the metabolic response was not robustly predictable from the *PAH* genotypes, based on the study design adopted for these clinical trials, and accordingly it seems prudent to test each person for this phenotype with a standardized protocol.

Communicated by: Nenad Blau.

Competing interests: None declared.

Electronic supplementary material The online version of this article (doi:10.1007/8904_2011_96) contains supplementary material, which is available to authorized users.

C.N. Sarkissian · P. Scott · C.R. Scriver
Departments of Biology, Human Genetics and Pediatrics, McGill University, Montreal, QC, Canada

C.N. Sarkissian · C.R. Scriver
Debelle Laboratory, McGill University-Montreal Children's Hospital Research Institute, 2300 Tupper Street, A-710b, Montreal, QC H3H 1P3, Canada

A. Gamez
Centro de Biología Molecular Severo Ochoa, Nicolas Cabrera 1 Laboratorio 204. Campus Cantoblanco, Universidad Autónoma de Madrid, 28049 Cantoblanco, Madrid, Spain

J. Dauvillier
Merck Serono S.A, 9 chemin des Mines, 1202, Geneva, Switzerland

A. Dorenbaum
BioMarin Pharmaceutical Inc, 105 Digital Drive, Novato, CA 94949, USA

R.C. Stevens (✉)
Department of Molecular Biology, The Scripps Research Institute, 10550 North Torrey Pines Road, La Jolla, CA 92037, USA
e-mail: stevens@scripps.edu
*These authors contributed equally to this work

Introduction

Phenylketonuria (PKU) is an autosomal recessive disorder affecting L-phenylalanine oxidation (OMIM #261600). In PKU, mutations in the phenylalanine hydroxylase (*PAH*) gene impair function of the hepatic enzyme (PAH; E.C. 1.14.16.1). Persistent hyperphenylalaninemia (HPA) has a neurotoxic effect resulting in mental retardation. The PKU/HPA phenotype can be ameliorated by restricting the dietary supply of the essential amino acid, phenylalanine (Phe) (Donlon et al. 2010).

Prevention of this form of mental retardation in the population (prevalence $\sim 10^{-4}$) begins with newborn screening for early detection of PKU/HPA followed by early and continuing treatment with a diet low in Phe. This

regimen became an epitome of human biochemical genetics (Donlon et al. 2010; Scriver 2007; Scriver and Clow 1980a, b). However, dietary treatment of PKU is difficult to maintain (Burgard et al. 1999) and is easily compromised, with three out of four PKU patients becoming noncompliant by late adolescence. Accordingly, discovery of a subset of PKU/HPA patients (Kure et al. 1999), who respond to daily pharmacological oral doses of tetrahydrobiopterin (BH_4), the catalytic cofactor for the PAH enzyme, has attracted interest (Bernegger and Blau 2002; Panel NIoHCD 2001; Phenylketonuria MRCWPo 1993). For such patients, treatment with oral BH_4 in pharmacological doses improves enzyme activity, enhances Phe oxidation, and leads to reduction in blood Phe levels (Burton et al. 2010; Elsas et al. 2011; Levy et al. 2007; Muntau et al. 2002).

The potential for a simple and effective oral therapy that improves control of blood Phe levels in PKU/HPA led to the development of sapropterin, a synthetic 6R-epimer of BH_4 [(containing $6R$)-L-*erythro*-5,6,7,8 tetrahydrobiopterin]. Safety and efficacy of sapropterin have been demonstrated in various studies including Phase II and III clinical trials (Burton et al. 2007; Lee et al. 2008; Levy et al. 2007; Trefz et al. 2009a), and other studies in selected populations have offered estimates of the BH_4-responsive phenotype (Blau and Trefz 2002; Dobrowolski et al. 2009a; Dobrowolski et al. 2011; Karacić et al. 2009; Lindner et al. 2001; Muntau et al. 2002; Spaapen and Rubio-Gozalbo 2003; Trefz et al. 2001; Trefz et al. 2010; Weglage et al. 2002). Here, we offer an estimate (17%) of the sapropterin-responsive phenotype in a large multi-national clinical study. *PAH* locus genotypes were also obtained in this study and did not reveal a robust genotype–phenotype correlation. Our findings have relevance for the management of PKU patients as well as indicating a need for improved protocol design to generate more comprehensive data sets in related clinical trials.

Methods

Clinical Trials Study Design, Measure of Efficacy and Analysis

Recruitment criteria and patient selection, test product (sapropterin dihydrochloride-available under the trade name KUVAN®) description, dose and mode of administration, study design and treatment schedules as well as method of serum [Phe] analysis were described in the clinical trials protocols (Burton et al. 2007; Lee et al. 2008; Levy et al. 2007). The three clinical trials (designated as PKU-001, PKU-003 and PKU-004) are registered with ClinicalTrials. gov.: NCT00104260, NCT00104247, NCT00225615. As part of the PKU-004 protocol, a population pharmacokinetics

analysis was performed, and some subjects participated in an adjunct open-label sub-study to evaluate the effect of once-daily dosing of sapropterin on Phe levels over a 24-h period (not reported by Lee et al. 2008).

The primary efficacy outcome variable was the metabolic response to sapropterin or placebo. It was defined as the percent change in blood Phe levels at a pre-determined observation point of treatment compared with baseline, prior to sapropterin treatment (for PKU-003, the baseline Phe level used was a calculated level using several pre-dose levels).

In screening study PKU-001, subjects who had a reduction in blood Phe level of at least 30% in study were defined as responders. If the observed response rate was 30%, the 95% confidence interval (CI) was expected to be 26–35% (Burton et al. 2007).

In the randomized placebo-controlled study PKU-003, the mean change in blood Phe levels, for 6 weeks, in each group was compared using an analysis of covariance model, with baseline Phe levels and treatment as the only covariates (Levy et al. 2007). The models utilized a last observation carried forward (LOCF) imputation approach to address missing data. The mean change in weekly blood Phe levels during the 6 weeks of treatment was evaluated using a longitudinal model. The proportion of subjects who had blood [Phe] $\leq$ 600 µM at week 6 was assessed using a Fisher's Exact Test.

In the safety and efficacy study PKU-004, blood Phe values for each subject, corresponding to the end of each 2-week dosing period (5, 20, and 10 mg/kg/day) over 6 weeks, were used to estimate the effect of dose on blood Phe concentration. Long-term persistence of response to sapropterin was measured up to week 22. The short-term effect of once-a-day dosing of sapropterin was measured by comparing blood Phe concentrations at set intervals over a 24-h period, in the PKU-004 sub-study (Lee et al. 2008). Analysis of variance for crossover designs was used to estimate average within-person changes in blood Phe concentration for the three dose levels. False positives, incorrectly assigned as responders in PKU-001, were identified using these findings.

PAH Gene Mutation Analysis

Informed consent was received for mutation analysis (conducted on DNA from blood samples collected in protocol PKU-001) or for the use of data obtained from archival clinical records documenting previous mutation analysis. Coded samples were submitted to the Molecular Genetics Laboratory of the Montreal Children's Hospital/ McGill University Health Centre for *PAH* gene sequencing. Genomic DNA was isolated from EDTA-blood from each participant using a blood kit (Qiagen). Polymerase chain reaction (PCR) amplification was performed on genomic

DNA for all exons and adjacent intronic splice regions. Amplicons were sequenced using BigDye terminator cycle sequencing kit (version 3.1) and run on the ABI 3100 Genetic Analyzer (Applied Biosystems). Sequence changes were detected using Mutation Explorer software (Version 2.41, SoftGenetics).

Primary Data

All data collected on participants, including site of study, race-ethnicity, mutation-genotype identification and measure of serum [Phe] throughout each of the clinical trials that the subject participated in, were recorded.

Allele or Genotype Relationships with Phenotype

PAH alleles were classified as (a) Responsive: when the mutation present in a homozygous and/or functionally hemizygous genotype was associated with responsiveness in PKU-001, in PKU-003 (if sapropterin was administered) and PKU-004. If a mutation occurring in subjects with homozygous genotypes was always associated with responsiveness, and yet some variation existed with hemizygous subjects carrying the mutation in question, this mutation was still considered responsive; (b) Ambiguously Responsive: when the mutation was associated with both responsiveness and unresponsiveness in different subjects with homozygous or equivalently hemizygous genotypes. A final verdict on responsiveness was not possible in individuals carrying mutations that did not fulfill the criteria of homozygosity or functional hemizygosity; (c) Unresponsive: when the mutation present in a homozygous and/or functionally hemizygous genotype was not associated with responsiveness in the clinical protocol PKU-001, or showed responsiveness in PKU-001 but not in PKU-003 (if sapropterin was administered) or PKU-004. Mutations that did not belong to any of the above categories and/or occurred only in a genotype carrying three mutations remained unclassified.

Genotypes were classified in a similar manner as either (a) Responsive: when they demonstrated responsiveness in all three clinical protocols; (b) Ambiguously responsive: when the same genotype was associated with a variation in responsiveness between subjects throughout the protocols; or (c) Unresponsive: when they were not associated with responsiveness in clinical protocol PKU-001 or demonstrated responsiveness in PKU-001 but not in PKU-003 (if sapropterin was administered) and/or PKU-004. Note: (a) If a genotype proved to be associated with responsiveness in PKU-001, and on subsequent evaluation the subject displayed a blood [Phe] reduction of $\geq 27\%$ in PKU-003 and/or PKU-004, the genotype was considered responsive. (b) Genotypes of participants that proved to be responsive or

ambiguously responsive only in PKU-001, and did not participate in further studies or participated only in the placebo series in PKU-003 and/or dropped out of the extended study (PKU-004) early before generating further comprehensive data, were accordingly assigned to the separate groups titled responsive or ambiguously responsive according to PKU-001 findings only. (c) One subject who participated in PKU-001, 003, and 004 had very low Phe levels that prevented proper interpretation of the change with sapropterin treatment and adequate classification of responsiveness. Although this subject was still classified as unresponsive, this observation was also identified in the results as an exception.

Results

Participants

Four hundred and eighty-five subjects completed protocol PKU-001 of which the majority ($>95\%$) were Caucasian; the remainder were of Hispanic, Arabic, Middle Eastern, African and Asian Pacific origins; 88 of the putative sapropterin-responsive group discovered in protocol PKU-001 then participated in PKU-003. Forty-seven of these subjects were assigned to the placebo group, and 41 to the sapropterin treatment group; 87 of these subjects actually completed the protocol. Eighty subjects from the PKU-003 cohort continued on to PKU-004; one noncompliant subject withdrew; 12 subjects joined the PKU-004 subgroup in which 24-h response to a single dose of sapropterin was evaluated.

Primary Data

The complete set of data for protocols PKU-001, 003, 004 and 004 (sub-study) were recorded and are available online in Supplementary Table 1.

Clinical Results

Ninety-six subjects (19.8% of cohort) from Protocol PKU-001 were responsive to sapropterin according to the criteria ($\geq 30\%$ decline in the level of blood Phe).

The randomized, double-blind, placebo-controlled trial (PKU-003) revealed that sapropterin was the effective therapeutic agent and the therapeutic effect was persistent, week by week.

Protocol PKU-004 showed a dosage response to sapropterin at three concentrations (5, 10, and 20 mg/kg/day) (Supplementary Table 2). There was an inverse relationship between drug dosage and blood Phe concentration. The average reduction of Phe levels for responders over the doses 5, 10, and 20 mg/kg/day was 16% (-156 μM), 28% (-265 μM), and 39% (-329 μM), respectively.

Supplementary Table 2 describes the responsive and unresponsive groups of patients participating in PKU-004. Of the 80 patients that participated in PKU-004, 69 responded to sapropterin treatment during the long-term protocol; the remaining 11 patients were identified as nonresponders; however, they were originally and falsely identified as responders in PKU-001. Thus, a more accurate estimation of the overall responders participating in these clinical trials is 17%.

The PKU-004 sub-study, which was designed to measure daily fluctuations in the middle of the long-term treatment, demonstrated stable blood Phe levels throughout the day. The fluctuations were minimal with an initial drop of approximately 16% 8 h postdosing and a stable and gradual rise in Phe levels by 16 h postdosing (12 midnight).

Mutation Analysis and Genotype–Phenotype Relationships

Responsiveness, based on data collected from PKU-004, identified false positives in patients participating in post-PKU-001 protocols. These findings were incorporated into the mutation and genotype assignments presented in Tables 1 and 2.

Among those enrolled in Protocol PKU-001, we obtained complete *PAH* genotypes (both alleles) in 424 subjects and partial genotypes (one allele) in 34 subjects; two patients carried three mutations. Among the fully genotyped subjects, only 61 (14.4%) harbored a homozygous genotype, while 363 subjects (85.6%) were compound heterozygotes.

We detected 122 different *PAH* mutant alleles (Table 1) harbored in 245 different genotypes. We classified the alleles as sapropterin-responsive ($n = 6$), ambiguous in their response (different response in different patients harboring the same allele) ($n = 10$), or unresponsive ($n = 73$) (Table 1). We could not classify 33 different alleles due to lack of homozygosity or pairing with a null allele. We mapped the point mutations and the associated state of responsiveness to sapropterin on to the molecular model of the PAH subunit (Fig. 1).

The most prevalent allele, identified in 145 of 458 completely or partially genotyped patients, was the null allele c.1222C>T (p.R408W). Ten alleles (Table 3) had not been previously reported in the literature or catalogued in the *PAH* gene mutation database (http://www.pahdb.mcgill.ca).

Among the 245 different genotypes (Table 2), 40 were responsive to sapropterin, 26 were ambiguously responsive and 179 were unresponsive.

Discussion

Successful dietary treatment of PKU/HPA due to deficient PAH enzyme function is now seen as a landmark and

paradigm shift in our overall view of human biochemical genetics (Donlon et al. 2010; Scriver 2007). The enzyme deficiency is now known to reflect a large array of mutant alleles and genotypes at the human *PAH* locus. The associated metabolic phenotype in some patients is ameliorated in response to pharmacologic oral doses of a synthetic *6R* epimer of tetrahydrobiopterin ($6R$-BH$_4$). The epimer is sapropterin dihydrochloride, marketed under the name KUVAN®. Many benefits would accrue to the patient if oral sapropterin could be integrated with dietary control (Blau et al. 2009); hence, our interest in knowing the prevalence of sapropterin-responsive PKU/HPA, and the likelihood of predicting a responsive or unresponsive phenotype from the associated genotype.

It is not surprising that many new reports, not all cited here, have appeared on this and related topics such as: the pharmacokinetic features of the oral agent; a defined metabolic response to the agent under a standardized protocol; whether dietary tolerance of Phe is improved in the responsive patient; and whether the agent is safe for long-term use and without adverse effects. However, it is already apparent that not all PKU/HPA patients can or will respond to oral sapropterin therapy. The responders are to some extent dependent on drug dosage, they tend to have more benign metabolic phenotypes in the untreated state, they are not equally distributed in human populations (Guldberg et al. 1998; Kayaalp et al. 1997; Lindner et al. 2003), and the distribution reflects that of mutant genotypes in the population. Moreover, the beneficial response to sapropterin is a complex person-specific process (Trefz et al. 2009b).

We examined responsiveness to sapropterin in a large international randomly selected group of PKU patients. Efficacy was measured by the decline in blood Phe concentration in the PKU patients after receiving sapropterin. Our analysis used the combined results of the three different components of the clinical trials, and draws conclusion on genotype-phenotype relationships based solely on the entire population that participated. It examines the associations of responsiveness with the *PAH* mutation and genotype.

Each protocol in the sapropterin project served a different purpose: Protocol PKU-001 yielded the apparent prevalence of sapropterin responsiveness in the sample (Burton et al. 2007); protocol PKU-003, confirmed that sapropterin is an agent conferring responsiveness (Levy et al. 2007); protocol PKU-004 revealed a dose-related response to sapropterin (Lee et al. 2008), showing that once-daily dosing is an effective regimen and provided evidence of false positive classifications in protocol PKU-001. All data for all subjects and all components of these clinical trials are available in Supplementary Table 1.

BH$_4$ is the catalytic cofactor for several enzymes, including PAH, and mutations exist that impair the enzymes

Table 1 Classification of mutant *PAH* alleles, in the present study, grouped by the phenotypic response to sapropterin (ordered 5′–3′)

Responsive

p.I65T	p.Q226H	p.A309D	
p.A104D	p.R241H	p.Y414C	

Ambiguously responsive

p.F39L	p.G218V	p.A345S	c.1315+1G>A
p.L48S	p.R261Q	p.L348V	
p.R68S	p.A309V	c.1066-3C>T	

Unresponsive

c.47_48delCT (p.S16>XfsX1)	c.503delA (p.Y168>Sfs)	p.R252W	c.969+5G>A
c.60+5G>T	p.G171R	p.L255S	p.Y325X
c.115_117delTTC (p.F39del)	p.R176X	p.A259V	p.G344R
p.S40L	p.V177L	p.R261X	p.S349P
c.165delT (p.F55>Lfs)	p.Y179N	p.R270G	p.G352R
p.F55L	p.Y179H	p.G272X	c.1066-11G>A
p.E56D	p.W187X	c.822_832del11 (p.K274_T278>Nfs)	p.Y356X
c.168+5G>C	p.C203W	p.Y277D	p.K371R
c.206_207insC (p.P69>Pfs)	p.Y204C	p.T278N	p.Y386C
p.T81P	p.Y204X	p.E280K	p.V388M
c.266_267insC (p.P89>Pfs)	c.632delC (p.P211>Hfs)	p.P281L	p.A395P
c.283_285delATC (p.I94del)	p.L213P	c.842+3G>C	c.1200-1G>A
p.R111X	p.Y216X	c.842+5G>A	p.R408W
c.398_401delATCA (p.N133_Q134>Rfs)	c.663_664delAG (p.E221_D222>Efs)	p.F299C	p.R408Q
c.441+2T>G	p.P225T	c.912+1G>A	p.A447D
c.441+5G>T	p.G239V	c.913-7A>G	c.1355_1356insA (p.K452>Kfs)
p.R158W	p.R243X	p.I306V	
p.R158Q	p.R243Q	p.S310F	
p.N167I	p.L249F	c.967_969delACA (p.T323del)	

Unclassified

p.Q20H	p.G188D	p.R243L	p.D338Y
p.I65N	p.V190A	p.R261P	p.Y343C
p.E78K	c.580_581delCT (p.L194>Efs)	p.T266P	c.1055delG (p.G352>Vfs)
p.D84Y	p.L194P	p.I269N[a]	p.Y386D
c.284_286delTCA (p.I95_K96delinsK)	p.P211T	c.806delT (p.I269>Tfs)	p.Y387H
c.441+1delGT	p.L212P	c.842+1G>A	p.E390G
c.441+1G>A	p.D222G	p.D282N	p.R413P
c.442-5C>G	p.S231P	c.919delG	
	p.R241C	p.Y325C	

[a] p.I269N was only present in a subject classified as having three mutations

serving synthesis and recycling of BH_4 (Thony and Blau 2006); these Mendelian disorders of BH_4 metabolism can be treated by sapropterin replacement therapy. The metabolic responsiveness of our subset of sapropterin-responsive HPA/PKU patients is explained by a different process, namely enhancement of residual PAH enzyme activity and opening of the oxidative pathway (Muntau et al. 2002). Pharmacological doses of sapropterin appear to achieve this effect, mainly in association with *PAH* missense alleles, either by a kinetic effect to overcome unfavorable binding of cofactor, or a chaperone-like effect on a misfolding enzyme subunit (Bechtluft et al. 2007; Dobrowolski et al.

Table 2 Classification of mutant *PAH* Genotypes, in the present study, grouped by the phenotypic response to sapropterin (ordered from 5′ to 3′ according to the mutation closest to the 5′)

Responsive

c.115_117delTTC (p.F39del): p.A345S	p.R68S: c.1066-11G>A	p.R243L: p.R261Q
p.F39L: p.I65T	p.R68S: c.1315+1G>A	c.806delT (p.I269>Tfs): p.Y414C
p.L48S: p.I65T	c.283_285delATC (p.I94del): p.R261Q	p.G272X: c.1066-3C>T
p.L48S: p.F299C	c.283_285delATC (p.I94del): p.P281L	p.Y277D: p.Y343C
p.L48S: p.S349P	p.A104D: p.A104D	p.E280K: p.L348V
p.L48S: p.Y387H	p.A104D: p.D222G	p.F299C: p.Y414C
c.165delT (p.F55>Lfs): p.Q226H	p.A104D: p.Y414C	p.A309D: p.R408W
p.I65T: p.I65T	p.P211T: p.R261Q	p.L348V: p.L348V
p.I65T: c.442-5C>G	p.G218V: p.P281L	c.1066-3C>T: c.1315+1G>A
p.I65T: p.Y277D	p.R241H: p.R252W	p.V388M: p.E390G
p.R68S: p.R252W	p.R241H: p.G272X	

Ambiguously responsive

p.F39L: p.R408W	p.I65T: p.R408W	p.G272X: p.Y414C
p.L48S: p.R261Q	p.I65T: p.Y414C	p.E280K: p.R408W
p.L48S: c.1066-11G>A	p.I65T: c.1315+1G>A	p.A309V: p.R408W
p.L48S: p.R408W	c.441+5G>T: p.R158Q	p.L348V: p.R408W
p.I65T: p.R243X	p.R243X: p.R408W	c.1066-11G>A: p.Y414C
p.I65T: p.P281L	p.R261Q: c.842+1G>A	p.R408W: p.Y414C
p.I65T: p.L348V	p.R261Q: c.1066-11G>A	c.1315+1G>A: c.1315+1G>A

Classified as responsive according to PKU-001 findings only

p.L48S: p.R158Q	p.E78K: p.R408W	p.A309V: p.Y414C
p.L48S: p.P281L	p.R158Q: p.R241C	p.E390G: p.R408W
p.I65T: p.R252W	p.R261Q: p.Y414C	

Classified as ambiguously responsive according to PKU-001 findings only

p.F39L: p.F39L	p.R261Q: p.R261Q	p.D282N: p.Y414
p.I65T: p.E280K	p.P281L: p.Y414C	

Unresponsive

c.47_48delCT (p.S16>XfsX1):c.47_48delCT (p.S16>XfsX1)	c.283_285delATC (p.I94del): c.1315+1G>A	p.R241H: p.R408W
c.47_48delCT (p.S16>XfsX1): p.A345S	p.A104D: c.822_832del11 (p.K274_T278>Nfs)	p.R243X: p.R243X
c.47_48delCT(p.S16>XfsX1)[a]: p.V388M[a]	p.A104D: p.R408W	p.R243X: p.Y414C
p.Q20H: p.R158Q	p.R111X: p.R111X	p.R243Q: p.R243Q
c.60+5G>T: p.R243Q	p.R111X: p.R408W	p.R243Q: p.R261Q
c.60+5G>T: c.1066-11G>A	p.R111X: p.R408Q	p.R243Q: p.L348V
c.60+5G>T: p.R408W	c.398_401delATCA (p.N133_Q134>Rfs): p.R408W	p.R243Q: p.R408W
c.60+5G>T: p.Y414C	c.441+1delGT: p.R158Q	p.L249F: p.L249F
c.60+5G>T: c.1315+1G>A	c.441+2T>G: p.R408W	p.L249F: p.R408W
c.115_117delTTC (p.F39del): c.822_832del11 (p.K274_T278>Nfs)	c.441+5G>T: p.V388M	p.R252W: p.P281L
c.115_117delTTC (p.F39del): p.S349P	p.R158W: p.R261P	p.R252W: c.1066-11G>A
p.F39L: p.R252W	p.R158W: p.R408W	p.R252W: c.1315+1G>A
p.F39L: p.E280K	p.R158Q: c.580_581delCT (p.L194>Efs)	p.L255S: p.L255S
p.F39L: c.1066-11G>A	p.R158Q: p.L212P	p.A259V: c.1066-11G>A
p.F39L: c.1315+1G>A	p.R158Q: p.L213P	p.R261X: p.R261X
p.S40L: c.842+1G>A	p.R158Q: p.R243X	p.R261X: p.Y325X
p.S40L: p.R408W	p.R158Q[a]: p.R261Q[a]	p.R261X: c.1066-11G>A

Table 2 (continued)

Unresponsive

p.L48S: p.I65N	p.R158Q: p.P281L	p.R261X: p.R408W
p.L48S: p.R111X	p.R158Q: p.F299C	p.R261P: p.Y414C
p.L48S: p.R252W	p.R158Q: c.1066-11G>A	p.R261Q[a,b]: p.G272X[a,b]
c.165delT (p.F55>Lfs): c.441+5G>T	p.R158Q: p.Y356X	p.R261Q: p.P281L
c.165delT (p.F55>Lfs): p.R158Q	p.R158Q: p.Y386D	p.R261Q: p.F299C
c.165delT (p.F55>Lfs): p.R252W	p.R158Q: p.V388M	p.R261Q: c.913-7A>G
c.165delT (p.F55>Lfs): p.R261X	p.R158Q: p.R408W	p.R261Q: p.L348V
c.165delT (p.F55>Lfs): p.R261Q	p.R158Q: p.Y414C	p.R261Q[a]: p.R408W[a]
c.165delT (p.F55>Lfs): p.G272X	p.R158Q: c.1315+1G>A	p.R261Q: c.1315+1G>A
c.165delT (p.F55>Lfs): p.P281L	p.N167I: p.G272X	p.R270G: p.R408W
c.165delT (p.F55>Lfs): p.R408W	c.503delA (p.Y168>Sfs): p.R408W	p.G272X: p.S349P
p.F55L: c.1200-1G>A	p.G171R: p.R408W	p.G272X: c.1066-11G>A
p.E56D: p.E56D	p.R176X: p.R176X	p.G272X: p.R408W
c.168+5G>C: c.168+5G>C	p.V177L: p.R408W	p.G272X: c.1315+1G>A
c.168+5G>C: p.I65T	p.Y179N: c.1066-11G>A	p.Y277D: c.1066-11G>A
c.168+5G>C: p.R408W	p.Y179H: p.R408W	p.Y277D[a]: c.1315+1G>A[a]
p.I65T[a]: p.R68S[a]	p.W187X: c.632delC (p.P211>Hfs)	p.T278N: p.P281L
p.I65T: p.E78K	p.V190A: p.T266P	p.T278N: c.1315+1G>A
p.I65T: c.441+5G>T	c.580_581delCT (p.L194>Efs): p.L348V	p.E280K: p.E280K
p.I65T: p.R158W	p.L194P: p.Y325C	p.E280K: p.P281L
p.I65T: p.R158Q	p.C203W: c.1315+1G>A	p.E280K: c.1066-11G>A
p.I65T: p.F299C	p.Y204C: c.663_664delAG (p.E221_D222>Efs)	p.E280K: c.1315+1G>A
p.I65T: c.912+1G>A	p.Y204X: p.Y414C	p.P281L: p.P281L
p.I65T: c.1066-11G>A	p.Y204X: c.1315+1G>A	p.P281L: p.S349P
p.I65T: p.R408Q	p.L213P: p.L213P	p.P281L: c.1066-11G>A
p.R68S: p.G272X	p.L213P: p.E280K	p.P281L: p.Y356X
p.R68S: p.R408W	p.L213P: p.R408W	p.P281L: p.R408W
c.206_207insC (p.P69>Pfs): c.1066-11G>A	p.Y216X: c.967_969delACA (p.T323del)	p.P281L: c.1315+1G>A
p.T81P: c.822_832del11 (p.K274_T278>Nfs)	p.G218V: p.S349P	c.842+1G>A: p.Y414C
p.T81P: p.R408W	c.663_664delAG (p.E221_D222>Efs): p.P281L	c.842+3G>C: p.R408W
p.D84Y: p.P281L	p.P225T: p.R243X	c.842+5G>A: p.Y414C
c.266_267insC (p.P89>Pfs): c.1066-11G>A	p.S231P: p.R261Q	p.F299C: c.1066-11G>A
c.283_285delATC (p.I94del): p.R408W	p.G239V: c.1066-11G>A	p.F299C: p.R408W
p.F299C: c.1315+1G>A	p.S349P: c.1315+1G>A	p.R408W[a]: p.R408W[a]
c.912+1G>A[a]: c.1055delG (p.G352>Vfs)[a]	p.G352R: c.1315+1G>A	p.R408W[a]: c.1315+1G>A[a]
c.912+1G>A: p.R408W	c.1066-11G>A: c.1066-11G>A	p.R408W: p.A447D
p.I306V: c.1066-11G>A	c.1066-11G>A: p.K371R	p.R408W: c.1355_1356insA (p.K452>Kfs)
p.S310F: c.1066-3C>T	c.1066-11G>A: p.R408W	p.R408W: p.L48S-p.I269N[c]
c.969+5G>A: c.969+5G>A	c.1066-11G>A: c.1315+1G>A	p.R408W: p.R413P-p.Y414C[c]
p.A345S: c.1066-11G>A	p.Y356X: p.Y414C	p.R408Q: c.1315+1G>A
p.L348V: c.1066-11G>A	p.Y386C[a]: c.1315+1G>A[a]	p.R413P: p.Y414C
p.L348V: c.1315+1G>A	p.A395P: p.R408W	p.Y414C: c.1315+1G>A
p.S349P: c.1066-11G>A	c.1200-1G>A: c.1315+1G>A	

[a] These genotypes were classified as unresponsive based on findings of PKU-001 and PKU-003 and/or PKU-004

[b] This genotype was classified as unresponsive according to the rules stated in the protocol; however, the levels were so low in the beginning of PKU-003 and PKU-004 that a true measure was not possible

[c] Double mutant; mutation is not completely classified

2009b; Erlandsen et al. 2004; Pey et al. 2007; Scavelli et al. 2005; Scriver and Waters 1999; Waters et al. 1999, 2000). Accordingly, it is relevant to know which among the several hundred *PAH* alleles (see www.pahdb.mcgill.ca) are associated with sapropterin responsiveness.

We classified the *PAH* mutant alleles (Table 1) and genotypes (Table 2), identified in the participants of the Phase II and III clinical trials, as either responsive, ambiguously responsive and unresponsive to sapropterin or unclassifiable, by using a simple metabolic (phenotypic) response as measured in these studies; while recognizing that other methods of response classification exist (Langenbeck 2008), which may influence the estimates of response prevalence. We used homozygous or hemizygous: null genotypes to classify the alleles, as was done earlier to describe PKU/HPA phenotypes (Guldberg et al. 1998; Kayaalp et al. 1997). We noticed that the proportion of homozygous genotypes in the present study (14.4%) is less than the frequency (~25%) observed previously in the earlier studies (Guldberg et al. 1998; Kayaalp et al. 1997), a finding that suggests selective sampling of the patient population in these studies. The most prevalent allele (p.R408W, c.1222C>T) is a nonresponsive null allele identified in 145 of 458 patients that were completely or partially genotyped; it reflects the predominance of subjects of northern European origins enrolled in the clinical study (http://www.pahdb.mcgill.ca). The frequency of sapropterin-responsive patients has varied between our own and other reports, a feature that could be explained in part by population genetics and the nonrandom distribution of mutant sapropterin-responsive *PAH* alleles (Guldberg et al. 1998; Kayaalp et al. 1997).

We mapped the responsive, ambiguously responsive and unresponsive point mutations onto the PAH crystal structure (Fig. 1a–c, respectively) according to the finding of these clinical trials. The responsive mutations seldom mapped to the cofactor binding site. One responsive allele, p.Y414C, maps to the dimer–dimer interfaces of the tetramer, suggesting that sapropterin enhances stabilization of the tetramer (Erlandsen et al. 2004; Pey et al. 2004). Among the other responsive mutations, p.I65T and p.A104D are located in the regulatory domain, while p.Q226H, p.R241H, and p.A309D are part of the catalytic domain (Fig. 1a). Of the mutations identified as ambiguously responsive, p.F39L, p.L48S, and p.R68S are located in the regulatory domain and p.G218V, p.R261Q, p.A309V, p.A345S, and p.L348V in the catalytic domain (Fig. 1b). Because our study is not a comprehensive sample of the human PKU/HPA population, a limitation compounded by a relaxed protocol design, we could not readily reveal the true prevalence of responsive mutations; for example, E390G in the catalytic domain is strongly associated with

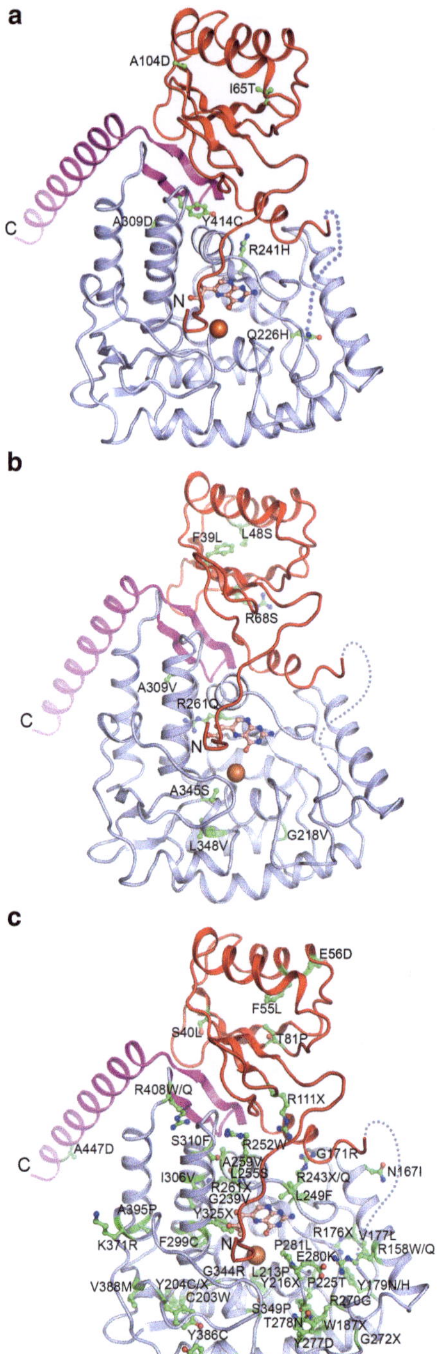

Fig. 1 Mapping of the identified point mutations on one enzyme subunit of the PAH crystal structure composite model (PDB ID code 1PAH): (**a**) responsive mutations, (**b**) ambiguously responsive mutations, and (**c**) unresponsive mutations

a chaperone-like effect (Erlandsen et al. 2004; Pey et al. 2004) yet did not readily disclose this property here.

Genotypes classified as responsive in these clinical trials have at least one mutation with residual PAH activity (see www.pahdb.mcgill.ca for data on activity of mutant alleles,

Table 3 Novel alleles not previously catalogued in www.pahdb. mcgill.ca

Mutation name	Nucleotide name	Responsiveness
p.Q20H	c.60G>C	Unclassified
p.P69>Pfs	c.206_207insC	Unresponsive
IVS4+1delGT	441+1delGT	Unclassified
IVS4+2T>G	c.441+2T>G	Unresponsive
p.Y179N	c.535T>A	Unresponsive
p.G188D	c.563G>A	Unclassified
p.C203W	c.609C>G	Unresponsive
p.T266P	c.796A>C	Unclassified
p.R270G	c.808A>G	Unresponsive
p.Y386D	c.1156T>G	Unclassified

e.g. the responsive allele p.R241H: paired with the null allele p.G272X). There was also unexpected responsiveness with genotypes classically associated with null alleles (e.g. p.G272X: c.1066-3C>T); and there were inconsistencies in the response of patients carrying previously characterized genotypes (e.g., p.I65T:p.R68S which were formerly described as responsive (Lindner et al. 2003), but in our study were unresponsive). Some of these anomalies may be explained by unmonitored dietary escape during clinical trials.

Genotypes bearing mutant alleles: p.I65T, p.A104D, p.Q226H, p.R241H, p.A309D, or p.Y414C (Table 1) were usually responsive to sapropterin (Table 2) in these clinical trials. However, these alleles were not necessarily predictive of responsiveness because genotypes harboring p.I65T, p.A104D, p.R241H, and p.Y414C were found in at least one sapropterin-unresponsive patient, leading us to conclude that factors other than *PAH* genotype contribute to sapropterin responsiveness. Inconsistencies in predicting phenotype from genotype is a recognized problem at the *PAH* locus (Guldberg et al. 1998; Kayaalp et al. 1997) and confounds earlier hopes to the contrary (Okano et al. 1991; Scriver 1991).

The response of certain PKU patients to BH_4 treatment has been postulated to occur through a variety of mechanisms (Erlandsen et al. 2004), one of which may involve overcoming suboptimal in vivo BH_4 concentrations (Kure et al. 2004). Consistent with this proposed mechanism, and other findings (Fiori et al. 2005; Hennermann et al. 2005; Matalon et al. 2004), we found that some putatively severe PKU phenotypes were partially responsive to sapropterin treatment in our clinical trials. We are aware that classification of responsiveness is dependent not only on the mutant allele at the *PAH* locus, but also on the genotype at that locus and on modifiers, yet to be identified, in the genome (Scriver and Waters 1999; Waters et al. 1999).

Other mechanisms to explain a positive or ambiguous sapropterin response may exist in mutations that affect splicing (Desviat et al. 2004) or modulate *PAH* gene expression (Blau and Trefz 2002). Splice-site mutations may generate multiple forms of transcripts and, thus, multiple forms of the encoded enzyme, a subset of which may be active. In this situation, positive response to sapropterin would depend on efficiency in generating a transcript encoding an enzyme with some activity. If the splice-site mutation increases the variability of the generation of specific transcripts, then the allele may show ambiguous or inconsistent response to sapropterin. In this study, the only intronic mutations classified as showing ambiguous sapropterin response were c.1066-3C>T and c.1315+1G>A, whereas others remained unresponsive or unclassifiable (Table 1).

Other factors that could contribute to an inconsistency in our findings could include once again the limited methodology with unmonitored changes in diet or other external factors such as intercurrent illness (Burton et al. 2007) and variability between individuals that may reflect sapropterin absorption and pharmacokinetics or bioavailability of the cofactor in the individual patient (Fiege et al. 2004; Leuzzi et al. 2006; Nielsen et al. 2010; Shintaku et al. 2005). Meanwhile, studying the residual PAH activity arising from the interaction of the mutant PAH subunits, carried by the PKU patients, is essential and may also influence BH_4-responsiveness. In this regard, a measure of predictability is enhanced when studying patients with homozygous or functionally hemizygous genotypes (Dobrowolski et al. 2009b, 2011; Fiori et al. 2005; Kure et al. 1999). The inconsistencies between genotype and BH_4/sapropterin response lead us to conclude that the nature of the response to BH_4 treatment is still not fully understood.

Our evidence that responsiveness to sapropterin is influenced by dosage of the agent (PKU-004) suggests a chemical response of the mutant protein in the presence of either a chaperone-like molecule or mass action to overcome impaired binding, as proposed elsewhere (Kure et al. 2004). The findings also highlight the individuality of the response to sapropterin therapy, the extent of which will be revealed in continuing studies that will measure, among others, the metabolic, cognitive, and psychological responses. While noting that our initial classification process for responsiveness in these clinical trials was based on response to sapropterin in the short-term protocol (PKU-001), we also recognize that there might be a different rate of responsiveness to be revealed when other more stringent short-term protocols and long-term exposures to sapropterin are considered. This also accounts for the minor inconsistencies of the rate of response reported here, as compared to what was reported in the prior and related clinical trial findings.

Once again, and in concordance with earlier findings of many others, we show that sapropterin responsiveness is not robustly predicted from genotype alone. In addition, though we acknowledge that the protocol parameters adapted in these clinical trials were designed to accommodate every day patient behavior, given the inconsistencies found between the clinical trial results reported here and other related studies, we recommend further review and application of more stringent and universally approved protocols and methodologies during upcoming drug trials. In the interim, we propose that it currently remains safer to identify the phenotype by direct observation of the response to sapropterin than to rely on prediction by genotype. Once again, we return to the thought that we should investigate and treat the patient and not just the genotype.

Acknowledgements We are indebted to the PKU patients and families who enrolled this study, as well as doctors and their healthcare staff for their invaluable assistance in the conduct of the clinical studies. We also thank our colleagues, John Tomaro for data collection, Sonia Schnieper-Samec for help with statistical review, Kumar Saikatendu and Katya Kadyshevskaya for the 3D figure preparation, Angela Walker for assistance with manuscript preparation and submission to the journal, Sun Sook Kim and Sabrina Cheng for data revision, and Manyphong Phommarinh and Jacques Mao for assistance with PAHdb. A. Gamez was supported by a research contract from "Ramón y Cajal" program by Ministerio de Ciencia e Innovación and Fundación Ramón Areces.

Synopsis

Some *PAH* mutations causing misfolding of the PAH protein respond to pharmacological doses of sapropterin dihydrochloride (BH_4) acting as a chaperone; however, to know genotype is not a robust predictor of therapeutic response, an assumption corroborated by findings reported in Phase II and III clinical trials.

Author Contributions

All co-authors participated in various aspects of the study. CNS and AG organized, corrected, analyzed, and interpreted the data, PS did the genotyping, JD completed the analysis of genotypes, AD compiled the data, CNS, AG and CRS drafted the manuscript, and CNS, AG, AD, RCS, and CRS reviewed it. The final version was seen and approved by all authors. CNS and AG are co-first authors who contributed equally to this work.

Guarantor

Raymond C. Stevens

Competing Interests Statement

The authors report commercial affiliations and competing financial interests: this study was supported by BioMarin Pharmaceutical Inc., the manufacturer of KUVAN®. AD was an employee of BioMarin Pharmaceutical Inc., and owns stock or stock options in the company. RCS and CRS have consulted (or are consultants) for BioMarin Pharmaceutical Inc. regarding their development of treatments for PKU/HPA.

	CNS	AG	PS	JD	AD	CRS	RCS
1. Have you in the past 5 years accepted the following from an organization that may in any way gain or lose financially from the results of your study or the conclusions of your review, editorial, or letter:							
Reimbursement for attending a symposium?	Yes	No	No	No	Yes	Yes	No
A fee for speaking or for organizing education?	No	No	No	No	Yes	No	No
Funds for research or for a member of staff?	Yes	No	No	No	Yes	Yes	Yes
A fee for consulting?	Yes	No	No	No	Yes	Yes	Yes
2. Have you in the past 5 years been employed by an organization that may in any way gain or lose financially from the results of your study or the conclusions of your review, editorial, or letter?	Yes	Yes	No	Yes	Yes	Yes	Yes
Do you hold any stocks or shares in such an organization?	No	No	No	No	Yes	No	No
3. Have you acted as an expert witness on the subject of your study, review, editorial, or letter?	No	No	No	No	No	No	No
4. Do you have any other competing financial interests?	No	No	No	No	No	No	No

Funding

This study grew out of clinical trials NCT00104260, NCT00104247, NCT00225615 supported by BioMarin Pharmaceutical Inc., the manufacturer of KUVAN®. The mutations analysis was funded by BioMarin Pharmaceutical Inc.; however, they did not financially support the authors (except for AD who was an employee at the time) in the organization, correction, analysis, and interpretation of the

data, drafting of the manuscript and review. BioMarin Pharmaceutical Inc. did not influence the analysis process or outcome of the project.

Ethics Approval

- Ethics approval for this research study was covered as a component of the clinical trials.
- Patient consent for this research study was covered as a component of the clinical trials.
- No vertebrate animals were used.

References

Bechtluft P, van Leeuwen RG, Tyreman M et al (2007) Direct observation of chaperone-induced changes in a protein folding pathway. Science 318(5855):1458–1461

Bernegger C, Blau N (2002) High frequency of tetrahydrobiopterin-responsiveness among hyperphenylalaninemias: a study of 1,919 patients observed from 1988 to 2002. Mol Genet Metab 77(4):304–313

Blau N, Trefz FK (2002) Tetrahydrobiopterin-responsive phenylalanine hydroxylase deficiency: possible regulation of gene expression in a patient with the homozygous L48S mutation. Mol Genet Metab 75(2):186–187

Blau N, Bélanger-Quintana A, Demirkol M et al (2009) Optimizing the use of sapropterin (BH(4)) in the management of phenylketonuria. Mol Genet Metab 96(4):158–163

Burgard P, Bremer HJ, Buhrdel P et al (1999) Rationale for the German recommendations for phenylalanine level control in phenylketonuria 1997. Eur J Pediatr 158(1):46–54

Burton BK, Grange DK, Milanowski A et al (2007) The response of patients with phenylketonuria and elevated serum phenylalanine to treatment with oral sapropterin dihydrochloride (6R-tetrahydrobiopterin): A phase II, multicentre, open-label, screening study. J Inherit Metab Dis 30(5):700–707

Burton BK, Bausell H, Katz R, Laduca H, Sullivan C (2010) Sapropterin therapy increases stability of blood phenylalanine levels in patients with BH4-responsive phenylketonuria (PKU). Mol Genet Metab 101(2–3):110–114

Desviat LR, Perez B, Belanger-Quintana A et al (2004) Tetrahydrobiopterin responsiveness: results of the BH4 loading test in 31 Spanish PKU patients and correlation with their genotype. Mol Genet Metab 83(1–2):157–162

Dobrowolski SF, Borski K, Ellingson CC, Koch R, Levy HL, Naylor EW (2009a) A limited spectrum of phenylalanine hydroxylase mutations is observed in phenylketonuria patients in western Poland and implications for treatment with 6R tetrahydrobiopterin. J Hum Genet 54(6):335–339

Dobrowolski SF, Pey AL, Koch R et al (2009b) Biochemical characterization of mutant phenylalanine hydroxylase enzymes and correlation with clinical presentation in hyperphenylalaninaemic patients. J Inherit Metab Dis 32(1):10–21

Dobrowolski SF, Heintz C, Miller T et al (2011) Molecular genetics and impact of residual in vitro phenylalanine hydroxylase activity on tetrahydrobiopterin responsiveness in Turkish PKU population. Mol Genet Metab 102(2):116–121

Donlon J, Levy HL, Scriver CR (2010) Hyperphenylalanine: phenylalanine hydroxylase deficiency. In: Scriver CR, Beaudet AL, Sly WS, Valle D, Childs B, Kinzler KW (eds) The metabolic and molecular bases of inherited diseases. McGraw-Hill, New York. Online. http://genetics.accessmedicine.com

Elsas LJ, Greto J, Wierenga A (2011) The effect of blood phenylalanine concentration on Kuvan™ response in phenylketonuria. Mol Genet Metab 102(4):407–412

Erlandsen H, Pey AL, Gamez A et al (2004) Correction of kinetic and stability defects by tetrahydrobiopterin in phenylketonuria patients with certain phenylalanine hydroxylase mutations. Proc Natl Acad Sci USA 101(48):16903–16908

Fiege B, Ballhausen D, Kierat L et al (2004) Plasma tetrahydrobiopterin and its pharmacokinetic following oral administration. Mol Genet Metab 81(1):45–51

Fiori L, Fiege B, Riva E, Giovannini M (2005) Incidence of BH4-responsiveness in phenylalanine-hydroxylase-deficient Italian patients. Mol Genet Metab 86(S1):S67–74

Guldberg P, Rey F, Zschocke J et al (1998) A European multicenter study of phenylalanine hydroxylase deficiency: classification of 105 mutations and a general system for genotype-based prediction of metabolic phenotype. Am J Hum Genet 63(1):71–79

Hennermann JB, Buhrer C, Blau N, Vetter B, Monch E (2005) Long-term treatment with tetrahydrobiopterin increases phenylalanine tolerance in children with severe phenotype of phenylketonuria. Mol Genet Metab 86(S1):S86–90

Karacić I, Meili D, Sarnavka V et al (2009) Genotype-predicted tetrahydrobiopterin (BH4)-responsiveness and molecular genetics in Croatian patients with phenylalanine hydroxylase (PAH) deficiency. Mol Genet Metab 97(3):165–171

Kayaalp E, Treacy E, Waters PJ, Byck S, Nowacki P, Scriver CR (1997) Human phenylalanine hydroxylase mutations and hyperphenylalaninemia phenotypes: a metanalysis of genotype-phenotype correlations. Am J Hum Genet 61(6):1309–1317

Kure S, Hou DC, Ohura T et al (1999) Tetrahydrobiopterin-responsive phenylalanine hydroxylase deficiency. J Pediatr 135(3):375–378

Kure S, Sato K, Fujii K et al (2004) Wild-type phenylalanine hydroxylase activity is enhanced by tetrahydrobiopterin supplementation in vivo: an implication for therapeutic basis of tetrahydrobiopterin-responsive phenylalanine hydroxylase deficiency. Mol Genet Metab 83(1–2):150–156

Langenbeck U (2008) Classifying tetrahydrobiopterin responsiveness in the hyperphenylalaninaemias. J Inherit Metab Dis 31(1):67–72

Lee P, Treacy EP, Crombez E et al (2008) Safety and efficacy of 22 weeks of treatment with sapropterin dihydrochloride in patients with phenylketonuria. Am J Med Genet A 146A(22):2851–2859

Leuzzi V, Carducci C, Chiarotti F, Artiola C, Giovanniello T, Antonozzi I (2006) The spectrum of phenylalanine variations under tetrahydrobiopterin load in subjects affected by phenylalanine hydroxylase deficiency. J Inherit Metab Dis 29(1):38–46

Levy HL, Milanowski A, Chakrapani A et al (2007) Efficacy of sapropterin dihydrochloride (tetrahydrobiopterin, 6R-BH4) for reduction of phenylalanine concentration in patients with phenylketonuria: a phase III randomised placebo-controlled study. Lancet 370(9586):504–510

Lindner M, Haas D, Mayatepek E, Zschocke J, Burgard P (2001) Tetrahydrobiopterin responsiveness in phenylketonuria differs between patients with the same genotype. Mol Genet Metab 73(1):104–106

Lindner M, Steinfeld R, Burgard P, Schulze A, Mayatepek E, Zschocke J (2003) Tetrahydrobiopterin sensitivity in German patients with mild phenylalanine hydroxylase deficiency. Hum Mutat 21(4):400

Matalon R, Koch R, Michals-Matalon K et al (2004) Biopterin responsive phenylalanine hydroxylase deficiency. Genet Med 6(1):27–32

Muntau AC, Roschinger W, Habich M et al (2002) Tetrahydrobiopterin as an alternative treatment for mild phenylketonuria. N Engl J Med 347(26):2122–2132

Nielsen JB, Nielsen KE, Güttler F (2010) Tetrahydrobiopterin responsiveness after extended loading test of 12 Danish PKU patients with the Y414C mutation. J Inherit Metab Dis 33 (1):9–16

Okano Y, Eisensmith RC, Guttler F et al (1991) Molecular basis of phenotypic heterogeneity in phenylketonuria. N Engl J Med 324 (18):1232–1238

Panel NIoHCD (2001) National Institutes of Health Consensus Development conference statement: Phenylketonuria: screening and management. Pediatrics 10:972–982

Pey AL, Perez B, Desviat LR et al (2004) Mechanisms underlying responsiveness to tetrahydrobiopterin in mild phenylketonuria mutations. Hum Mutat 24(5):388–399

Pey AL, Stricher F, Serrano L, Martinez A (2007) Predicted effects of missense mutations on native-state stability account for phenotypic outcome in phenylketonuria, a paradigm of misfolding diseases. Am J Hum Genet 81(5):1006–1024

Phenylketonuria MRCWPo (1993) Phenylketonuria due to phenylalanine hydroxylase deficiency: an unfolding story. Br Med J 306 (6870):115–119

Scavelli R, Ding Z, Blau N, Haavik J, Martinez A, Thony B (2005) Stimulation of hepatic phenylalanine hydroxylase activity but not Pah-mRNA expression upon oral loading of tetrahydrobiopterin in normal mice. Mol Genet Metab 86(S1):S153–155

Scriver CR (1991) Phenylketonuria – genotypes and phenotypes. N Engl J Med 324(18):1280–1281

Scriver CR (2007) The PAH gene, phenylketonuria, and a paradigm shift. Hum Mutat 28(9):831–845

Scriver CR, Clow CL (1980a) Phenylketonuria: epitome of human biochemical genetics (first of two parts). N Engl J Med 303 (23):1336–1342

Scriver CR, Clow CL (1980b) Phenylketonuria: epitome of human biochemical genetics (second of two parts). N Engl J Med 303 (24):1394–1400

Scriver CR, Waters PJ (1999) Monogenic traits are not simple: lessons from phenylketonuria. Trends Genet 15(7):267–272

Shintaku H, Fujioka H, Sawada Y, Asada M, Yamano T (2005) Plasma biopterin levels and tetrahydrobiopterin responsiveness. Mol Genet Metab 86(S1):S104–106

Spaapen LJ, Rubio-Gozalbo ME (2003) Tetrahydrobiopterin-responsive phenylalanine hydroxylase deficiency, state of the art. Mol Genet Metab 78(2):93–99

Thony B, Blau N (2006) Mutations in the BH4-metabolizing genes GTP cyclohydrolase I, 6-pyruvoyl-tetrahydropterin synthase, sepiapterin reductase, carbinolamine-4a-dehydratase, and dihydropteridine reductase. Hum Mutat 27(9):870–878

Trefz FK, Aulela-Scholz C, Blau N (2001) Successful treatment of phenylketonuria with tetrahydrobiopterin. Eur J Pediatr 160(5):315

Trefz FK, Burton BK, Longo N et al (2009a) Efficacy of sapropterin dihydrochloride in increasing phenylalanine tolerance in children with phenylketonuria: a phase III, randomized, double-blind, placebo-controlled study. J Pediatr 154(5):700–707

Trefz FK, Scheible D, Götz H, Frauendienst-Egger G (2009b) Significance of genotype in tetrahydrobiopterin-responsive phenylketonuria. J Inherit Metab Dis 32(1):22–26

Trefz FK, Scheible D, Frauendienst-Egger G (2010) Long-term follow-up of patients with phenylketonuria receiving tetrahydrobiopterin treatment. J Inherit Metab Dis Mar 9 Epub

Waters PJ, Parniak MA, Akerman BR, Jones AO, Scriver CR (1999) Missense mutations in the phenylalanine hydroxylase gene (PAH) can cause accelerated proteolytic turnover of PAH enzyme: a mechanism underlying phenylketonuria. J Inherit Metab Dis 22(3):208–212

Waters PJ, Parniak MA, Akerman BR, Scriver CR (2000) Characterization of phenylketonuria missense substitutions, distant from the phenylalanine hydroxylase active site, illustrates a paradigm for mechanism and potential modulation of phenotype. Mol Genet Metab 69(2):101–110

Weglage J, Grenzebach M, von Teeffelen-Heithoff A et al (2002) Tetrahydrobiopterin responsiveness in a large series of phenylketonuria patients. J Inherit Metab Dis 25(4):321–322

JIMD Reports
DOI 10.1007/8904_2011_99

Riboflavin-Responsive Trimethylaminuria in a Patient with Homocystinuria on Betaine Therapy

Nigel J. Manning · Elizabeth K. Allen ·
Richard J. Kirk · Mark J. Sharrard · Edwin J. Smith

Received: 10 August 2011 /Revised: 10 August 2011 /Accepted: 23 September 2011 /Published online: 20 November 2011
© SSIEM and Springer-Verlag Berlin Heidelberg 2011

Abstract A 17-year-old female patient with pyridoxine non-responsive homocystinuria, treated with 20 g of betaine per day, developed a strong body odour, which was described as fish-like. Urinary trimethylamine (TMA) was measured and found to be markedly increased. DNA mutation analysis revealed homozygosity for a common allelic variant in the gene coding for the TMA oxidising enzyme FMO3. Without changing diet or betaine therapy, riboflavin was given at a dose of 200 mg per day. An immediate improvement in her odour was noticed by her friends and family and urinary TMA was noted to be greatly reduced, although still above the normal range.

Gradual further reductions in TMA (and odour) have followed whilst receiving riboflavin. Throughout this period, betaine compliance has been demonstrated by the measurement of dimethylglycine (DMG) excretion, which has been consistently increased. Marked excretions of DMG when the odour had subsided also demonstrate that DMG was not the source of the odour.

This patient study raises the possibility that betaine may be converted to TMA by intestinal flora to some degree, resulting in a significant fish odour when oxidation of TMA is compromised by *FMO3* variants. The possibility exists that the body odour occasionally associated with betaine therapy for homocystinuria may not be related to increased circulating betaine or DMG, but due to a common FMO3 mutation resulting in TMAU. Benefits of riboflavin therapy for TMAU for such patients would allow the maintenance of betaine therapy without problematic body odour.

Introduction

Trimethylaminuria (TMAU) and its associated body odour ("Fish Odour Syndrome") can be caused by lack of trimethylamine (TMA) N-oxidation by the hepatic enzyme flavin containing mono-oxygenase type 3 (FMO3) (Primary TMAU) (MIM 136132) (Humbert et al. 1970) (Lee et al. 1976) as well as excess TMA production by intestinal flora (Secondary TMAU) (Mitchell 1996) (Fraser-Andrews et al. 2003). More than 40 mutations have been described in the *FMO3* gene, which is located on chromosome 1, with clear genotype–phenotype correlation (Treacy et al. 1998; Zschocke et al. 2002) including combinations of allelic variants, which have been associated with mild or intermittent TMAU. Most common is the variant p.[Glu158Lys; Glu308Gly] which, in homozygous state, may affect between 1 in 100 and 1 in 20 of the population (Zschocke et al. 1999). Presentation may be limited to periods of high dietary choline and seafood as well as peri-menstrual periods in female patients.

Body odour has been reported as a side effect of betaine (trimethylglycine) administration (Wilcken and Wilcken 1997) when used in therapeutic doses for the treatment of pyridoxine non-responsive homocystinuria (cystathionine beta-synthase deficiency (MIM 236200). The description of

Communicated by: Ertan Mayatepek

Competing interests: None declared.

N.J. Manning (✉) · E.J. Smith
Department of Clinical Chemistry, Sheffield Children's Hospital, Western Bank, Sheffield S10 2TH, UK
e-mail: Nigel.Manning@sch.nhs.uk

E.K. Allen · R.J. Kirk
Sheffield Diagnostic Genetics Service, Sheffield Children's Hospital, Sheffield, UK

M.J. Sharrard
Department of Paediatrics, Sheffield Children's Hospital, Sheffield, UK

the odour has been reported as fishy (Kraus and Kozich 2001) suggesting TMA, but could also be due to the demethylated metabolite of betaine, dimethylglycine (DMG). DMG has not only been reported at high concentrations in plasma and urine following betaine administration (Schwahn et al. 2003), but has been associated with a fish-like odour in a single case of DMG-dehydrogenase deficiency (Moolenaar et al. 1999).

Riboflavin (vitamin B2) administration has been associated with reduction of TMA excretion in some TMAU patients (unpublished observations) presumed to be due to increased FMO3 activity with riboflavin acting as a cofactor. Riboflavin responsiveness has previously been reported for some patients with multiple acyl-CoA dehydrogenase deficiency where riboflavin is a co-factor for the electron transfer flavoprotein-CoQ, which is vital for the activity of acyl-CoA dehydrogenases (Olsen et al. 2007). The case of TMAU presented here clearly demonstrates a response to riboflavin in lowering TMA excretion with concomitant reduction in body odour in a teenage girl receiving high-dose betaine therapy for homocystinuria.

Methods

TMA: Urinary measurement of TMA (with TMA-oxide by titanium chloride reduction) was achieved using alkalinised samples heated in a headspace autosampler with gas chromatography–mass spectrometry (GCMS) of gaseous TMA. Stable isotope ratio of TMA to deuterated (d9-) TMA was used for quantitation (Treacy et al. 1995).

40 μg of d9-TMA-DCl internal standard was added to 2 ml of urine for TMA (free) analysis and 0.2 ml of urine for free TMA plus TMA-oxide (total) analysis. Following reaction with titanium chloride for the total TMA measurement duplicates of free and total TMA for each urine were cooled in ice and alkalinised by addition of 0.6 g of KOH and 1 g of K_2CO_3. Calibration standards of TMA hydrochloride and TMA-oxide were taken through the method together with quality control samples at two levels.

Prepared samples were sealed in vials and loaded on to a headspace autosampler (HP7694 Agilent Technologies), which was linked through a heated line to a GCMS (Agilent 5973N) fitted with a 15 m fused silica column with no stationary phase. Samples were heated to 90°C for 40 min prior to injection. GCMS run time was 5 min during which TMA was monitored using the M-H ion (m/z 58) and d9-TMA with the M-D ion (m/z 68). Peak area ratios for both free and total TMA were used to calculate concentrations following calibration of the internal standard.

DMG: Urine with d2-DMG internal standard was analysed by lyophilisation and GCMS of tert-butyldimethylsilyl ester.

100 μl aliquots of urine (diluted to achieve a creatinine value of 0.1 mmol/L) were mixed with an aqueous solution containing 0.5 μg d2-DMG internal standard and 1 ml of methanol was added to aid lyophilisation. Vials were placed under a gentle stream of nitrogen for 45 min at 40°C. Dried samples were than derivatised by addition of 100 μl N-methyl-tert-butyldimethylsilyl-trifluoroacetamide (MTBSTFA) with 100 μl acetonitrile and heating to 80°C for 45 min. Tertiary butyl dimethylsilyl esters were injected on to a GC column (SGE BPX5 30m) using a 3 μl split injection. Quantitation was achieved by monitoring ions of m/z 160 and 162 for DMG and d2-DMG, respectively.

DNA: Patient genomic DNA was extracted from peripheral blood samples using Magnetic Separation Module I (Chemagen). Genomic DNA was amplified by PCR using Red Hot Taq polymerase (ABGene) with 3 mM $MgCl_2$.

Primers were designed to amplify each protein-coding exon and at least 25 bp of the intron/exon boundaries for exons 2–9 of the *FMO3* gene (Accession number NM_006894.5). Cycle sequencing was performed using standard M13 primers attached to the gene-specific primers with electrophoresis carried out on ABI3730 DNA analysers.

Case Report

SW was initially diagnosed with homocystinuria at 7 years of age after presenting with dislocated lenses. Her plasma total homocysteine (tHcys) was measured at 145 μmol/L (ref. < 16), which was partially responsive to pyridoxine (121 μmol/L). Further reduction of tHcys (59 μmol/L) was achieved by betaine therapy initially at three doses of 2 g per day, which was increased to two doses of 8 g per day by the time she was 15 years of age. Plasma methionine increased dramatically with betaine therapy from 57 to 1,445 μmol/L (ref. 8–47) indicating successful remethylation of homocystine. During teenage years plasma tHcys concentrations of more than 100 μmol/L gave cause for concern as linear growth came to an end, thus decreasing protein requirement and increasing protein catabolism. As a response to this change, betaine dosage was increased to 20 g per day, which unfortunately resulted in a lack of compliance (unused betaine was discovered at home) with resultant very high plasma tHcys results. Following clinical advice and family intervention, compliance improved and tHcys values started to normalise. At this new high dosage, however, a strong fishy body and breath odour was noticed by family and friends, which began to cause problems at school. The odour persisted when the betaine dose was reduced to 16 g per day. At 17 years of age, this serious social problem posed another threat to betaine compliance

Table 1 Urinary TMA and DMG μmol/mmol creatinine. Plasma DMG μmol/L before and after riboflavin administration

Days on riboflavin	Free TMA (ref. < 11)	Free/total TMA (ref. < 0.21)	DMG (ref. < 16)	Plasma DMG (ref. < 8)
0	392	0.93	1,471	272
28	77	0.36	1,076	216
77	37	0.27	1,017	
133	27	0.21	2,044	
223	22	0.17	1,935	
328	20	0.07	957	

and metabolic control of SW's homocystinuria. It was also reported that the odour was more noticeable around the time of menstruation. The family reported that the odour was directly related to betaine administration.

TMA was measured in a urine sample and found to be markedly increased at 392 mmol/mol creatinine (normal range 2–11). Her free TMA/total TMA (free TMA + TMA-oxide) ratio was also increased at 93% (normal range < 21%).

It was decided to not withdraw or modify betaine therapy or make any changes to her diet, but to try riboflavin at two doses of 100 mg per day.

Within days of riboflavin supplementation, SW's body odour had significantly improved and her family could hardly detect any body odour.

Urinary TMA was measured after 30 days and was found to have decreased to 77 (mmol/mol creat.) (Table 1) (36% free/total TMA). Further measurements of TMA showed continued reduction of excretion to 20 after 330 days on riboflavin (7% free/total TMA). SW has reported a residual fishy taste (breath odour only), however body odour is no longer a problem.

Urinary DMG measurements (Table 1) show the expected marked excretion consistent with betaine compliance ranging from 1.0 to 2.2 mol/mol creat. Plasma DMG was also measured and found to be 272 and 216 μmol/L (ref. < 8) at the time of the first two urines collected. Importantly, these DMG values in our patient were maintained after the body odour subsided.

Mutation Analysis

Genomic DNA analysis by sequencing revealed a previously described common variant allele p.Glu158Lys; Glu308Gly in the homozygous state (Zschocke et al. 1999). Previous reports have associated this genotype with a mild or intermittent phenotype, most likely to only present with a significant odour when challenged by dietary

TMA precursors or during hormonal fluctuation such as just prior to menstruation.

Discussion

Side effects of betaine (Cystadane) have been well documented and include anorexia, agitation, depression, irritability, personality disorder, sleep disturbance, dental disorders, diarrhoea, glossitis, nausea, stomach discomfort, vomiting, urinary incontinence, hair loss, hives and abnormal skin odour (Orphan-Europe 2007). These undesirable effects have been classified as "uncommon" with a frequency of between 1 in 100 and 1 in 1,000 patients.

In a pilot study of betaine therapy for non-alcoholic steatohepatitis, however, four of the ten subjects experienced nausea, abdominal cramps, loose stools and body odour (Abdelmalek et al. 2001).

Choline (2-hydroxy-trimethylammonio-ethanol) has been shown to be readily converted to TMA-oxide when fed to rats (Norris and Benoit 1945). In experiments comparing urinary TMA-oxide produced after injection and feeding approximately 27% of fed choline was converted to TMA-oxide compared to 2% by injection. This bacterial route was not reproduced by betaine feeding, which resulted in less than 1% conversion. Previously, intestinal bacterial production of TMA from betaine had been described as "trace" (Wunsche 1940). Therefore, it is unsurprising that betaine has been disregarded as a source of TMA by enterobacterial action (Busby et al. 2004).

Conversion of betaine to TMA as part of methanogenesis has been reported for some bacterial species, notably *Clostridium sporogenes* (Naumann et al. 1983), *Desulfuromonas acetoxidans* (Heijthuijsen and Hansen 1989), and *Haloanaerobacter salinarus* (Moune et al. 1999). Whether the dynamics of the pathways for these species could result in significant output of TMA, however, is not certain.

Our patient had no problematic side effects of betaine therapy from the age of 7 until 15 years. Her odour can be strictly correlated with TMAU and follows the course of the mild or intermittent form of this disorder, which is associated with the common allelic variant previously described. Female patients with this variant often suffer from a fish odour around the time of menstruation. The odour may then subside unless challenged with a significant dietary load (usually high-choline foods such as eggs, legumes, offal as well as seafood (TMA-oxide). Dietary loading with high-choline foods and marine fish has become a vital diagnostic tool, especially given that intermittent odour is a feature of some *FMO3* variants (Chalmers et al. 2003).

Our patient's presentation appears, therefore, to be strongly linked to the combined effects of betaine loading and a mild FMO3 deficiency caused by an underlying genetic condition other than her homocystinuria. The immediate reduction in TMA excretion when given 100 mg per day riboflavin without reducing betaine therapy demonstrates a mild phenotype, which would possibly present intermittently if loading with betaine was not ongoing.

Betaine compliance was demonstrated by a marked excretion of DMG, which remained elevated when the odour (and TMA excretion) was significantly reduced. Our patient's odour remained barely detectable following the establishment of riboflavin supplementation, although DMG excretion actually increased between 77 and 133 days after riboflavin. Plasma values were similar to those of the patient described with a fishy odour attributed to dimethyl-glycinuria (Moolenaar et al. 1999) with urinary values for our patient two- to fourfold greater than those of the reported dimethylglycinuria patient. Fish odour associated with increased DMG therefore appears to be inconsistent when taking into account our patient and other conditions, which result in increased DMG such as multiple acyl-CoA dehydrogenase deficiency (MADD) where no reports of odour have been cited (Burns et al. 1998) with excretion of DMG as marked as in the single reported case of DMG-dehydrogenase deficiency.

SW has continued to be betaine compliant even though a residual breath odour has been reported by her as the taste of fish. TMA excretion has continued to fall over the months since commencement of riboflavin and may eventually reach normal values. Without riboflavin administration, there is no doubt that taking high doses of betaine would have become difficult for this patient. Variable betaine compliance has been reported as an issue for the treatment of pyridoxine non-responsive homocystinuria (Singh et al. 2004). Side effects such as odour may have contributed to this lack of compliance and led to sub-optimal control of patients' metabolic state. Despite the routine listing of unusual body odour as an occasional side effect of betaine therapy (Orphan Europe 2007), the precise nature of the odour has not been described other than that of "fishy" (Kraus and Kozich 2001). Hypermethioninaemia from remethylation of homocystine by betaine has been associated not only with organ toxicity in rats and growth retardation in infants (Smolin et al. 1981) but also with a sulphurous, cabbage-like body odour (Påby et al. 1989). Therefore, a combination of odours may trouble some patients receiving betaine, both TMA and methionine contributing. Measurement of TMA, FMO3 genotyping and riboflavin therapy for some patients with homocystinuria may enable odour relief and promote compliance with high dosage betaine therapy and its life-saving benefits.

Synopsis

Betaine-related body odour may be due to trimethylaminuria and respond to riboflavin therapy without lowering betaine dosage.

References

Abdelmalek MF, Angulo P, Jorgensen RA et al (2001) Betaine, a promising new agent for patients with nonalcoholic steatohepatitis: results of a pilot study. Am J Gastroenterol 96:2711–2717

Burns SP, Holmes HC, Chalmers RA, Johnson A, Isles RA (1998) Proton NMR spectroscopic analysis of multiple acyl-CoA dehydrogenase deficiency – capacity of the choline oxidation pathway for methylation in vivo. Biochim Biophys Acta 1406:274–282

Busby MG, Fischer L, Da Costa KA et al (2004) Choline- and betaine- defined diets for use in clinical research and for the management of trimethylaminuria. J Am Diet Assoc 104:1836–1845

Chalmers RA, Bain MD, Iles RA (2003) Diagnosis of trimethylaminuria in children: marine fish versus choline load test. J Inherit Metab Dis 26 (Suppl 2):(448-P) 224

Fraser-Andrews EA, Manning NJ, Ashton GHS, Eldridge P, McGrath JA, Menagé H (2003) Fish odour syndrome with features of both primary and secondary trimethylaminuria. Clin Exp Dermatol 28:203–205

Heijthuijsen JHFG, Hansen TA (1989) Betaine fermentation and oxidation by marine *Desulfuromonas* strains. Appl Environ Microbiol 55(4):965–969

Humbert JR, Hammond KB, Hathaway WE, Marcoux JG, O'Brien D (1970) Trimethylaminuria: the fish-odour syndrome. Lancet 2:770–771

Kraus JP, Kozich V (2001) Cystathione beta-synthase and its deficiency. Homocysteine in Health and Disease. Carmel R, Jacobsen DW (eds) Cambridge University Press Chapt.20: 223–243

Lee WG, Yu JS, Turner BB, Murray KE (1976) Trimethylaminuria: fishy odors in children. N Engl J Med 295:937–938

Mitchell SC (1996) The fish odour syndrome. Perspect Biol Med 39 (4):514–526

Moolenaar SH, Poggi-Bach J, Engelke UFH, Cortisiaensen JMB et al (1999) Defect in dimethylglycine dehydrogenase, a new inborn error of metabolism: NMR spectroscopic study'. Clin Chem 45 (4):459–464

Moune S, Hirschler N, Caumette P, Willison J, Matheron R (1999) *Haloanaerobacter salinarius* a novel halophilic fermentative bacterium that reduces glycine-betaine to trimethylamine with hydrogen or serine as electron donors; emendation of the genus *Haloanaerobacter*. Int J Syst Bacteriol 49:103–112

Naumann E, Hippe H, Gottschalk G (1983) Betaine: new oxidant in the stickland reaction and methanogenesis from betaine and L-alanine by a *Clostridium sporogenes-Methanosarcina barkeri* co-culture. Appl Environ Microbiol 45(2):474–483

Norris ER, Benoit GJ (1945) Studies on trimethylamine oxide III. Trimethylamine oxide excretion by the rat. J Biol Chem 158:443–448

Olsen RKJ, Olpin SE, Andresen BS et al. (2007) ETFDH mutations as a major cause of riboflavin- responsive multiple acyl-CoA dehydrogenation deficiency. Brain 130:2045–2054

Orphan Europe (2007) Cystadane (betaine anhydrous). Adjunctive treatment for homocystinuria. CYSTAD 07-03-6645 (Brochure)

Påby P, Ekelund S, Bøgeskov-Jensen I (1989) A case of intermittent hypermethioninemia as the cause of disagreeable body odour (in Danish). Ugeskr Laeger 151(24):1551–1552

Singh RH, Kruger WD, Wang L, Pasquali M, Elsas LJ (2004) Cystathionine beta-synthase deficiency: effects of betaine supplementation after methionine restriction in B6-nonresponsive homocystinuria. Genet Med 6(2):90–95

Schwahn BC, Hafner D, Hohlfeld T, Balkenhol N, Laryea MD, Wendel U (2003) Pharmacokinetics of oral betaine in healthy subjects and patients with homocystinuria. Br J Clin Pharmacol 55: 6–13

Smolin LA, Benevenga NJ, Berlow S (1981) The use of betaine for the treatment of homocystinuria. J Pediatr 99:467–472

Treacy EP, Ackerman BR, Chow LML et al (1998) Mutations of the flavin- containing monooxygenase gene (FMO3) cause trimethylaminuria. A defect in detoxication. Hum Mol Genet 7:839–845

Treacy E, Johnson D, Pitt JJ, Danks DM (1995) Trimethylaminuria, fish odour syndrome: a new method of detection and response to treatment with metronidazole. J Inherit Metab Dis 18:306–312

Wilcken DEL, Wilcken B (1997) The long term outcome in homocystinuria. Rosenberg IH, Graham I, Ueland P et al. (eds): International Conference on Homocystineine Metabolism: From Basic Science to Clinical Medicine. Norwell, MA, Kluwer Academic: 51

Wunsche R (1940) Uber die Wirkung von Darmbakterien auf Trimethylammoniumbasen. Ber Ges Physiol Exp Pharm 118:303

Zschocke J, Kohlmueller D, Quak E, Meissner T, Hoffmann GF, Mayatepek E (1999) Mild trimethylaminuria caused by common variants in FMO3 gene. Lancet 354(9181):834–835

Zschocke J, Guldberg P, Wevers RA et al (2002) Molecular studies in flavin-containing mono-oxygenase 3 (FMO3) deficiency (abstract). J Inherit Metab Dis 25(Suppl 1):168

JIMD Reports
DOI 10.1007/8904_2011_100

Successful Noninvasive Ventilation and Enzyme Replacement Therapy in an Adult Patient with Morbus Hunter

M. Westhoff · P. Litterst

Received: 19 July 2011 /Revised: 09 September 2011 /Accepted: 04 October 2011 /Published online: 16 December 2011
© SSIEM and Springer-Verlag Berlin Heidelberg 2011

Abstract M. Hunter is characterized by an accumulation of mucopolysaccharides in cells, blood, and connective tissue as a consequence of a deficiency of the enzyme iduronate-2-sulfatase. Unlike enzyme replacement therapy with idursulfase in children, there is limited long-term experience in adult patients with Morbus Hunter.

The case presented here describes the development of a man born in 1971 who was admitted to Hemer Lung Clinic in 2005 with severe obstructive sleep apnea, pulmonary functional impairment, and ventilatory failure (FEV 1: 0.8 L, VC: 1.0 L; pO_2: 52 mmHg; pCO_2: 81 mmHg, 6 MWT: 100 m). Initially, the patient received symptomatic treatment with noninvasive ventilation, which achieved a considerable improvement in pulmonary function and a normalization of blood gasses. Since February 2008, the patient received additional enzyme replacement therapy with idursulfase, which resulted in a further significant functional improvement (FEV1: 1.6; VC: 2.3 L; VO_2max: 1,350 mL or 28.1 mL/kg body weight), in a normalization of prior elevated pulmonary artery pressures and also in impressive changes in the physiognomy and joint mobility. In November 2010, the polysomnography and nocturnal blood gas analysis without NIV showed only a mild obstructive sleep-related breathing disorder with no sign of hypoventilation. Therapy was changed to nocturnal CPAP therapy with a constant pressure of 6 cm H_2O. Additional administration of oxygen was not required. With this therapy, the patient has been asymptomatic up to September 2011.

Adult Hunter patients also benefit from enzyme replacement therapy and, in restrictive ventilatory defects with hypoventilation, from symptomatic therapy with noninvasive ventilation.

Abbreviations

6-MWT	Six-minute walk test
AHI	Apnea/hypopnea index
BGA	Blood gas analysis
CPAP	Continuous positive airway pressure
DLCO	Diffusing capacity of the lung for carbon monoxide
FEV1	Forced expiratory volume in 1 s
FVC	Forced vital capacity
GAG	Glycosaminoglycan
IVS	Inter ventricular septum
MPS	Mucopolysaccharidosis
NIV	Noninvasive ventilation
P01	Negative airway pressure
PA syst.	Pulmonary artery systolic pressure
pCO_2	Partial pressure of carbon dioxide
Pimax	Maximal inspiration pressure
pO_2	Partial pressure of oxygen
TLC	Total lung capacity
VC	Vital capacity
VO_2max	Maximum oxygen uptake

Communicated by: Verena Peters.

Competing interests: None declared.

Shire provided funding for translation service. The views in the manuscript do not necessarily reflect those of Shire.

M. Westhoff (✉) · P. Litterst
Department of Pneumology, Hemer Lung Clinic, Hemer 58675, Germany
e-mail: Michael.Westhoff@lkhemer.de

Introduction

Hunter syndrome belongs to the group of mucopolysaccharidoses (MPS) and is characterized by increased accumulation in the tissues and urinary excretion of glycosaminoglycans (GAG), in particular dermatan sulfate and heparan sulfate (Burrow and Leslie 2008; Martin et al. 2008). MPS II is an X-linked trait, so that the disease affects almost exclusively boys and men (Neufeld and Muenzer 2001; Wraith et al. 2008). The genetic defect results in an iduronate-2-sulfatase deficiency, the key enzyme for degradation of GAG (Le Guern et al. 1990; Burrow and Leslie 2008).

The disease is very rare, with an incidence of 0.64 in 100,000 births and 1.3 in 100,000 male births in Germany (Baehner et al. 2005). Typical changes in the physiognomy occur in childhood, with coarseness in the facial features resulting from the accumulation of GAG: nose with flattened bridge, thick lips, large head, prominent forehead, short neck, widely spaced teeth, as well as frizzy hair and thickened skin. In addition, other characteristic changes or diseases as otitis, hernia, diseased heart valves, hepatosplenomegaly, joint stiffness and kyphoscoliosis are often observed (Beck 2011; Martin et al. 2008; Jones et al. 2009; Young and Harper 1982, 1983). Respiratory symptoms are typical and a common cause of clinical deterioration particularly as the disease progresses and in adulthood. The reason for this is the accumulation of GAG in the soft tissue, which causes enlargement of the tonsils and tongue and also narrowing of the upper respiratory tract (Kurihara et al. 1992; Kamin 2008). In some cases, this results in obstructive sleep-related breathing disorders even in childhood (Shapiro et al. 1985; Ginzburg et al. 1990; Leighton et al. 2001) and also difficult intubation (Kamin 2008; Muenzer et al. 2009). In addition, disease progression with tracheobronchomalacia (Morehead and Parsons 1993) and tracheal stenosis in childhood and also in adulthood (Young and Harper 1979; Davitt et al. 2002; Gross and Lemmens 2010) have been described.

Until now the life expectancy of patients with Hunter syndrome has been low, especially if the disease was the severe form (Young and Harper 1983). Attenuated cases with almost normal life expectancy have also been described, although very rarely (Young and Harper 1982; Jones et al. 2009). Only around 10% of all Hunter patients reached the age of 25 before the introduction of enzyme replacement therapy (Wraith et al. 2008; Jones et al. 2009).

Enzyme replacement therapy, which has been available in Europe since 2007 in the form of idursulfase, represents a starting point for potentially preventing the hitherto typical clinical course of the disease in young patients who die before reaching adulthood (Fenton and Rogers 2006; Young and Harper 1983). Study results and initial experience of therapy with idursulfase suggest that enzyme replacement therapy should be begun early (Wraith et al. 2008) before irreversible changes have manifested and ideally before significant progression of the disease.

However, there is little experience to date with enzyme replacement therapy in adult Hunter patients, who show a classic manifestation of the clinical picture. The current publication presents the clinical progression of an adult Hunter patient who was initially treated with noninvasive ventilation (NIV) and has additionally been receiving enzyme replacement therapy since its marketing authorization.

Case Report

The male patient was born in January 1971. A possible storage disease was suspected at an early age, although there was no evidence to verify this. A blind hepatic biopsy was carried out in September 1974 in response to suspected glycogenosis, although the result was inconclusive. In October 1974, there was evidence of increased urinary excretion of acid mucopolysaccharides and MPS II was diagnosed.

The patient is not intellectually compromised and, upon leaving school, he undertook vocational training and worked at a computer workstation. As he got older, he developed dyspnea on exertion, which manifested even when walking normally at ground level.

In October 2005, the patient was admitted to Hemer Lung Clinic for further diagnosis and therapy. At this time, he had severe restrictive ventilatory problems with an FEV 1 of 0.8 L (26% of target), a VC of 1.0 L, and a TLC of 2.6 L (53% of target) (Table 1). Blood gas analysis (BGA) showed respiratory global insufficiency with a pH of 7.38, a pO_2 of 52 mmHg, and pCO_2 of 81 mmHg. The pulmonary artery pressure was raised to a systolic reading of 55 mmHg. The interventricular septum (IVS) was also thickened to 1.30 cm, and there was a dynamic left ventricular outflow tract obstruction with a pressure gradient of 60 mmHg and grade II aortic valve insufficiency. Urinary excretion of acid mucopolysaccharides was elevated at 5.0 mg/0.1 g creatinine (norm < 2.8). Nocturnal polysomnography detected a disturbed sleep profile with reduced deep sleep and REM sleep. The average oxygen saturation was 75% and the respiratory disturbance index was raised at 60/h. Nocturnal BGA showed respiratory acidosis with a pH of 7.30. The pCO_2 was 81 mmHg and the pO2 was 32 mmHg.

Treatment was initiated with gradual continuous positive airway pressure (CPAP) titration, adjusted to an ultimate

Table 1 Lung function and echocardiography prior to noninvasive ventilation therapy (10/2005), up to introduction of enzyme replacement therapy (2/2008) and after introduction of enzyme replacement therapy (9/2008)

Function parameter	Dates of assessments		
	10/2005	2/2008	9/2008
FEV1 (L)	0.8	1.5	1.6
VC (L)	1.0	2.0	2.1
TLC (L)	2.6	3.6	4.0
DLCO (% of target)		73.4	70.8
P0.1 (kPa)	0.59	0.47	0.30
Pimax (kPa)	3.42	4.56	4.23
pO2 (mmHg)	52	80	77
pCO2 (mmHg)	81	34	32
6MWT (m)	100(pO2 44 (mmHg)/ pCO2 66 (mmHg)	280(pO2 68 (mmHg)/ pCO2 43 (mmHg)	400(pO2 76 (mmHg)/ pCO2 38 (mmHg)
PA syst.	55	45	30
IVS thickness	1.3	1.15	1.05

pressure of 11 cm H_2O to open up the upper respiratory tract at night. However, hypercapnea persisted in the course of CPAP titration, sometimes with pCO_2 readings up to 110 mmHg and a pH of 7.22. Treatment was therefore definitively adjusted to NIV with a pressure of 31/11 cm H_2O, a frequency of 18/min with the additional administration of 2 L O2/min. This gradually achieved normalization of blood gas values, and the patient's sleep quality improved considerably.

NIV was subsequently used consistently. Pulmonary function results showed a clear improvement in February 2008 (Table 1). The FEV1 increased to 1.5 L, the VC to 2.0 L and the TLC to 3.6 L. Measurement of muscle power and respiratory pump load showed under NIV an increase in the Pimax from 3.43 kPa to 4.56 kPa and a simultaneous fall in the P0.1 reading from 0.59 kPa to 0.47 kPa. The blood gas readings at rest were pO_2 80 mmHg and pCO_2 34 mmHg. The distance covered in the Six-Minute Walk Test (6MWT) increased from 100 m in 2005 to 280 m in February 2008. The blood gas readings in the 6MWT test likewise showed a considerable improvement. Whereas in 2005 after walking 100 m, the pO_2 was found to decrease to 44 mmHg with an increase in pCO_2 to 66 mmHg, the pO_2 was 68 mmHg and the pCO_2 was 43 mmHg after walking a distance of 280 m (Table 1). The pulmonary artery systolic pressure was recorded as 45 mmHg.

In February 2008, supplementary therapy was initiated with idursulfase in a well-tolerated dosage of 0.5 mg/kg body weight. Under this treatment, the excretion of GAG was regularly monitored and found to have fallen to below half the upper limit of normal (58 µg/mg creatinine).

In September 2008, under idursulfase therapy an increase of the FEV1 to 1.6 L, of the VC to 2.1 L, and of the TLC to 4.0 L was detected, with a decrease in the P0.1 to 0.3 kPa and constant Pimax (Table 1). The BGA showed a tendency to hyperventilation (pO_2: 77 mmHg, pCO_2: 32 mmHg). The pulmonary artery pressure was normalized. Nocturnal polysomnography showed only a mild obstructive sleep-related breathing disorder. The nocturnal BGA readings were in the normal range. The NIV was maintained, and the therapeutic pressure was reduced to 20/9 cm H_2O. Administration of oxygen was no longer required.

In the six-minute walk test (6MWT), the distance covered had increased to 400 m. The BGA under exertion again showed an improvement (pO_2: 76 mmHg, pCO_2: 38 mmHg). At this point, spiroergometry was carried out for the first time, when a load of 80 W and a maximum oxygen uptake of 1,350 mL or 28.1 mL/kg body weight were achieved. The anaerobic threshold was 57% of the target maximum oxygen uptake. There were no indications of disturbed diffusion at an alveolar-arterial oxygen tension difference ($AaDO_2$) of 31 mmHg at maximum load.

The most recent follow-up examination in November 2010 under ongoing idursulfase therapy again showed an increase in the FEV1 to 1.6 L, of the TLC to 4.4 L, and of the VC to 2.3 L. The thickness of the septum had reduced further to 0.95 cm. Even if there have been only rather minor changes in the functional parameters from September 2008 to November 2010, a steady and clearly positive change has been recorded with regard to the joint and soft tissue situation and in particular to the physiognomy over the period of time in question. The joint contractures improved, especially in the area of the distal interphalangeal joints of the finger, so that the flexion which was previously fixed at 110° decreased to 85°, while the facial contours became smoother and the wrinkle formation in the face was reduced (Fig. 1). The previously frizzy hair appeared considerably straighter.

Polysomnography confirmed that the NIV was well adjusted, with a tendency to nocturnal hyperventilation and a consequent tendency to central apneas. The polysomnography without NIV showed only a mild obstructive sleep-related breathing disorder with no sign of nocturnal hypoventilation. Therefore, treatment was changed to nocturnal CPAP therapy with a constant pressure of 6 cm H_2O. Additional administration of oxygen was not required. With this therapy, the patient has been asymptomatic up to September 2011.

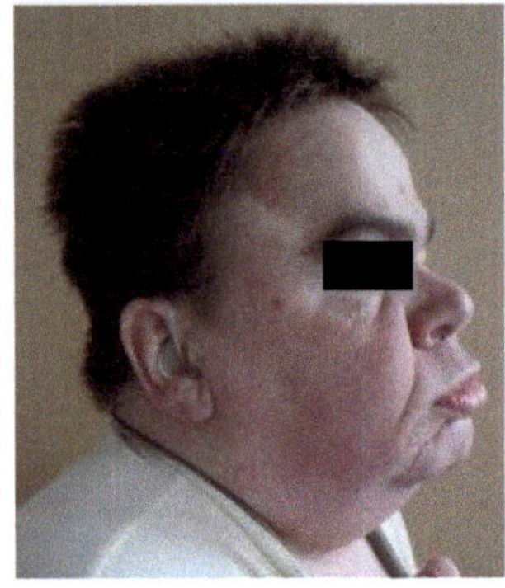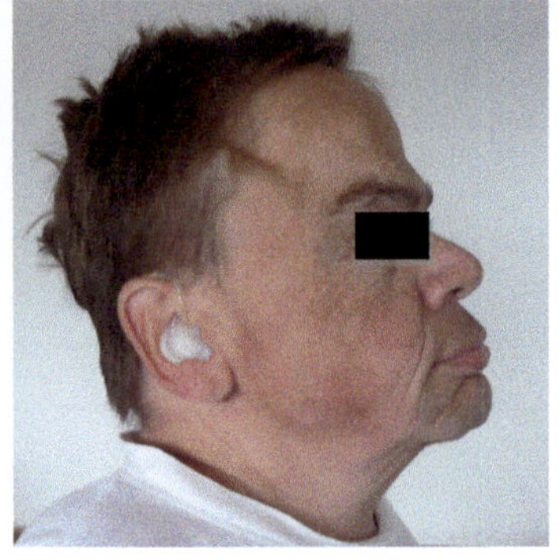

Fig. 1 Facial physiognomy of Hunter patient – in 2005 prior to idursulfase therapy (*left*) and in September 2011 after 3 years of idursulfase therapy (*right*)

Discussion

Until now only isolated case descriptions and clinical progression reports have been available for adult patients because of the typical progression of the disease – very few patients survive beyond the age of 30 (Pérez-Calvo et al. 2011). Most patients with Hunter syndrome have respiratory problems at an early age as a result of the progressive obstruction of the upper respiratory tract caused by an accumulation of GAG and restricted mobility of the mandibular joint. Some of these respiratory problems are associated with severe nocturnal sleep-related breathing disorders and frequent infections of the upper respiratory tract (Leighton et al. 2001; Orliaguet et al. 1999; Shapiro et al. 1985).

In many patients, these changes occur in the first or second decade of life, in particular when the clinical progression of the disease is severe, so that respiratory deterioration is a common cause of death (Young and Harper 1983). However, this type of severe obstruction of the upper respiratory tract is also described for attenuated forms of the disease, and this obstruction has been shown to determine the clinical progression (Young and Harper 1982). In isolated cases, there is also relevant thoracic restriction (Young and Harper 1982), as was also seen in the patient under discussion: in 2005, besides a disturbed sleep profile and sleep apnea, he also showed a severe restrictive ventilation disorder with hypoventilation.

In keeping with previous positive experiences in patients with Hunter syndrome (Leighton et al. 2001; Ginzburg et al. 1990), CPAP titration was primarily carried out to resolve the respiratory tract obstruction. This required a high CPAP level, although in comparative terms it was below previously reported CPAP levels of up to 20 cm H_2O (Ginzburg et al. 1990). Orliaguet et al. report an intolerance to CPAP pressures > 12 cm H_2O with the occurrence of dyspnea in their patients (Orliaguet et al. 1999). The cause was found to be a myxoma in the upper respiratory tract;

therefore, surgical resection of the myxoma in the area of the vocal cords was carried out. Only then, with the CPAP adjusted to 12 cm H_2O, were the AHI and the pCO_2 successfully normalized. Myxomatous changes of this type were not present in the case presented here, as well as other causes of obstruction described in the literature (Gross and Lemmens 2010; Young and Harper 1979; Davitt et al. 2002; Morehead and Parsons 1993), such as tracheal stenosis or tracheobronchomalacia. Although CPAP therapy effectively resolved the upper airway obstruction in the patient under discussion here, contrary to the case report of Ginzburg et al. in 1990, it did not remedy the accompanying hypoventilation, which had to be attributed to the severe restriction. However, the diurnal pCO_2 readings reported by Ginzburg et al. of a maximum of 51 mmHg were considerably lower.

Surgical therapy was not indicated due to the lack of evidence of surgically resectable myxomatous tissue. Therefore, the persistent hypoventilation with restrictive ventilatory disturbance and pronounced respiratory muscle weakness (raised p0.1 and lowered Pimax) presented an indication for initiating NIV. Blood gas levels were successfully normalized in the course of the NIV only after high inspiration pressures and a longer adaptation phase were selected. This avoided the need for a tracheotomy, as has been described in isolated cases with the most severe respiratory tract obstructions (Sasaki et al. 1987; Yoskovith et al. 1998), particularly as tracheotomy would also have been associated with an increased operative risk.

In contrast to CPAP therapy in Hunter syndrome with respiratory tract obstruction or sleep apnea, the use of NIV is not documented in the literature. Besides the rarity of these types of processes in adulthood, this is because the application of NIV has only become more widespread in the last two decades. Therefore, the functional improvements achieved under NIV in the course of 3 years cannot be compared to a similarly treated group. At any rate it is clear that, particularly in comparison with the functional improvements under the enzyme substitution initiated in February 2008, effective ventilation alone can achieve a considerable increase in lung volume, with a virtual doubling of the FEV1 and VC, and an increase of 1 L in the total lung capacity (TLC). The normalization of blood gas levels at rest and on exertion as well as the increase in the distance covered in the 6MWT are also to be attributed to the effect of relieving the burden on the respiratory muscles and to the demonstrable increase in respiratory muscle strength in the course of the disease as a result of the systematic application of NIV.

In contrast to the central sleep apnea in untreated patients with Hunter syndrome described by Kurihara et al. as a consequence of the accumulation of gangliosides in the respiratory center (Kurihara et al. 1992), the central apneas seen under NIV are to be interpreted as a consequence of hyperventilation with a drop in the pCO_2 below the apnea threshold and are thus an indicator of effective therapy.

The adult patient described here has been receiving a combination of NIV therapy and enzyme replacement therapy with once weekly 0.5 mg/kg idursulfase since February 2008. The improvement achieved in the symptoms, but also the persistence of the functional improvements initially achieved by the NIV after ending the NIV are clearly attributable to the enzyme replacement. After a first six-month check-up under enzyme replacement therapy and in the further clinical course until September 2011, accompanied by normalization of the urinary excretion of GAG, there was further improvement in the lung function values FEV1, VC, and TLC and the BGA under exertion, normalization of nocturnal hypoventilation and clear improvement in the obstructive sleep-related breathing disorder.

The functional improvements in lung function and in the load tests are considerably greater than the increases in lung volume and distances walked reported to date in therapeutic studies with idursulfase. The Japanese study (Okuyama et al. 2010) reported an increase in the distance walked of 54.5 m and in the FVC of 3.8% after 12 months, comparable with the data specified by Muenzer et al. 2006. They saw after 1 year of idursulfase therapy an increase of 37 m in the 6MWT ($P = 0.013$), a percentage increase in the predicted FVC of 2.7% ($P = 0.065$) and an increase in the absolute FVC of 160 mL ($P = 0.001$), compared to a 30%-increase (i.e., 120 m) in the 6MWT seen in our patient.

Since the functional improvements achieved primarily by the NIV also persisted after changing from NIV to CPAP therapy under enzyme substitution, there is much to suggest that these improvements could also have been achieved by enzyme replacement therapy alone, if it had been available at the time.

Likewise the normalization of the left ventricular wall thickness and the pulmonary artery pressure are attributable to enzyme replacement therapy, even if a clear reduction in right ventricular load and left ventricular hypertrophy had already been recorded under NIV.

Nevertheless, this is an unusual clinical improvement, especially since it occurred in an adult Hunter patient with long-established disease and severe respiratory impairment, in whom it would not have been primarily expected in this way before starting enzyme replacement therapy.

The increase in joint mobility which is impressive in a patient with long-standing disease duration was described in a similar way by Pérez-Calvo et al. in a 31-year-old male patient (Pérez-Calvo et al. 2011) and emphasizes that joint mobility can be improved by enzyme replacement therapy even in older patients, which has a significant effect on the use of the fingers in everyday life. The pronounced change and rejuvenation of the facial features is unusual for the age of the patient and is another indication of success of enzyme replacement therapy.

To sum up, the case presented here demonstrates that adult Hunter patients can benefit from enzyme replacement therapy, and in the case of a restrictive ventilatory disturbance with hypoventilation can benefit from symptomatic therapy with NIV. These treatments in adults can potentially achieve considerable functional improvements as well as changes in the physiognomy and joint mobility.

References

Baehner F, Schmiedeskamp C, Krummenauer F et al (2005) Cumulative incidence rates of the mucopolysaccharidoses in Germany. J Inherit Metab Dis 28:1011–1017

Beck M (2011) Mucopolysaccharidosis Type II (Hunter Syndrome): clinical picture and treatment. Curr Pharm Biotechnol 12:861–866

Burrow TA, Leslie ND (2008) Review of the use of idursulfase in the treatment of mucopolysaccharidosis II. Biologics 2:311–320

Davitt SM, Hatrick A, Sabharwal T, Pearce A, Gleeson M, Adam A (2002) Tracheobronchial stent insertions in the management of major airway obstruction in a patient with Hunter syndrome (type II mucopolysaccharidosis). Eur Radiol 12:458–462

Ginzburg AS, Onal E, Aronson RM, Schild JA, Mafee MF, Lopata M (1990) Successful use of nasal-CPAP for obstructive sleep apnea in Hunter syndrome with diffuse airway involvement. Chest 97:1496–1498

Gross ER, Lemmens HJ (2010) Hunter syndrome in an adult: beware of tracheal stenosis. Anesth Analg 110:642–643

Jones SA, AlmÄssy Z, Beck M et al (2009) Mortality and cause of death in mucopolysaccharidosis type II-a historical review based on data from the Hunter Outcome Survey (HOS). J Inherit Metab Dis 32:534–543

Kamin W (2008) Diagnosis and management of respiratory involvement in Hunter syndrome. Acta Paediatr 97:57–60

Kurihara M, Kumagai K, Goto K, Imai M, Yagishita S (1992) Severe type Hunter's syndrome. Polysomnographic and neuropathological study. Neuropediatrics 23:248–256

Le Guern E, Couillin P, Oberlé I, Ravise N, Boue J (1990) More precise localization of the gene for Hunter syndrome. Genomics 7:358–362

Leighton SE, Papsin B, Vellodi A, Dinwiddie R, Lane R (2001) Disordered breathing during sleep in patients with mucopolysaccharidoses. Int J Pediatr Otorhinolaryngol 58:127–138

Martin R, Beck M, Eng C et al (2008) Recognition and diagnosis of mucopolysaccaridosis II (Hunter syndrome). Pediatrics 121: e377–e386

Morehead JM, Parsons DS (1993) Tracheobronchomalacia in Hunter's syndrome. Int J Pediatr Otorhinolaryngol 26:255–261

Muenzer J, Wraith JE, Beck M et al (2006) A phase II/III clinical study of enzyme replacement therapy with idursulfase in mucopolysaccharidosis II (Hunter syndrome). Genet Med 8:465–473

Muenzer J, Beck M, Eng CM et al (2009) Multidisciplinary management of Hunter syndrome. Pediatrics 124:e1228–e1239

Neufeld EF, Muenzer J (2001) The mucopolysaccharidoses. In: Scriver CR, Beaudet AL, Sly WS et al (eds) The metabolic and molecular bases of inherited disease, 8th edn. Mc Graw-Hill, New York, pp 3421–3452

Okuyama T, Tanaka A, Suzuki Y et al (2010) Japan Elaprase Treatment (JET) study: idursulfase enzyme replacement therapy in adult patients with attenuated Hunter syndrome (Mucopolysaccharidosis II, MPS II). Mol Genet Metab 99:18–25

Orliaguet O, Pépin JL, Veale D, Kelkel E, Pinel N, Lévy P (1999) Hunter's syndrome and associated sleep apnoea cured by CPAP and surgery. Eur Respir J 13:1195–1197

Pérez-Calvo J, Bergua Sanclemente I, López Moreno MJ et al (2011) Early response to idursulfase in a 31-year old male patient with Hunter syndrome. Rev Clin Esp 211:e42–e45

Sasaki CT, Ruiz R, Gaito R Jr, Kirchner JA, Seshi B (1987) Hunter's syndrome: a study in airway obstruction. Laryngoscope 97:280–285

Shapiro J, Strome M, Crocker AC (1985) Airway obstruction and sleep apnea in Hurler and Hunter syndromes. Ann Otol Rhinol Laryngol 94:458–461

Wraith JE, Scarpa M, Beck M et al (2008) Mucopolysaccaridosis type II (Hunter syndrome): a clinical review and recommendations for treatment in the era of enzyme replacement therapy. Eur J Pediatr 167:267–277

Yoskovith A, Tewfik TL, Brouillette RT, Schloss MD, Der Kaloustian VM (1998) Acute airway obstruction in Hunter syndrome. Int J Pediatr Otorhinolaryngol 44:273–278

Young ID, Harper PS (1979) Long-term complications in Hunter's syndrome. Clin Genet 16:125–132

Young ID, Harper PS (1982) Mild form of Hunter's syndrome: clinical delineation based on 31 cases. Arch Dis Child 57:828–836

Young ID, Harper PS (1983) The natural history of the severe form of Hunter's syndrome: a study based on 52 cases. Dev Med Child Neurol 25:481–489

JIMD Reports
DOI 10.1007/8904_2011_101

RESEARCH REPORT

Hyperargininemia: A Family with a Novel Mutation in an Unexpected Site

Y. Haimi Cohen* · R. Bargal* · M. Zeigler ·
T. Markus-Eidlitz · V. Zuri · A. Zeharia

Received: 10 June 2011 / Revised: 6 October 2011 / Accepted: 10 November 2011 / Published online: 21 December 2011
© SSIEM and Springer-Verlag Berlin Heidelberg 2011

Abstract Hyperargininemia is a rare autosomal recessive disorder of the last step of the urea cycle characterized by a deficiency in liver arginase1. Clinically, it differs from other urea cycle defects by a progressive paraparesis of the lower limbs (spasticity and contractures) with hyperreflexia, neurodevelopmental delay and regression in early childhood. Growth is affected as well. Hyperammonemia is episodic, if present at all. The disease is caused by mutations in the *ARG1* gene; there are approximately 20 different known *ARG1* mutations with considerable genetic heterogeneity. We describe two Arab siblings with a late diagnosis of hyperargininemia and present the genetic findings in their family. As *ARG1* sequencing was unrevealing despite suggestive clinical and laboratory findings, molecular cDNA analysis was performed. The *ARG1* expression pattern identified a 125-bp out-of-frame insertion between exons 3 and 4, leading to the addition of 41 amino acids and a premature termination codon TGA at the sixth codon downstream. The insertion originated at intron 3 and was attributable to a novel c.305 + 1323 t > c intronic base change that enabled an enhancement phenomenon. This is the first reported exon-splicing-enhancer mutation in patients with hyperargininemia. The clinical course and genetic findings emphasize the possibility that hyperargininemia causes neurological deterioration in children and the importance of analyzing the expression pattern of the candidate gene when sequencing at the DNA level is unrevealing.

Communicated by: verena peters

Competing interests: None declared.

Y. Haimi Cohen (✉) · T. Markus-Eidlitz · A. Zeharia
Day Hospitalization Unit, Schneider Children's Medical Center of Israel, Petah Tiqwa 49202, Israel
e-mail: yhaimi@bezeqint.net

Y. Haimi Cohen · T. Markus-Eidlitz · A. Zeharia
Sackler Faculty of Medicine, Tel Aviv University, Tel Aviv, Israel

R. Bargal · M. Zeigler · V. Zuri
Department of Human Genetics and Metabolic Diseases,
Hadassah Hebrew University Hospital, Jerusalem, Israel
*Dr Y. Haimi-Cohen and R. Bargal contributed equally to this work as first authors

Hyperargininemia (OMIM 207800) is a rare autosomal recessive disorder in the last step of the urea cycle characterized by a deficiency in liver arginase1 (EC 3.5.3.1), which normally hydrolyzes arginine into ornithine. Hyperargininemia is caused by mutations in the 8 exons of the *ARG1* gene located on chromosome 6q23 (Sparkes et al. 1986). It is distinguished from other urea cycle disorders by the time of onset and the characteristic clinical picture. Hyperargininemia usually appears in infants and toddlers and rarely in the neonatal period (De Deyn et al. 1997). It manifests as progressive spastic paraparesis (less on the upper extremities), loss of developmental milestones which gradually evolves into severe mental retardation, poor growth with consequent short stature, and seizures; some patients experience episodes of irritability, nausea, poor appetite, and lethargy (Crombez and Cederbaum 2005). In neonates, hyperargininemia has been reported to present with cirrhosis (Braga et al. 1997), cholestasis (Gomes Martins et al. 2011), and cerebral edema (Picker et al. 2003). The disease usually develops insidiously; in rare cases, ammonia may rise to levels that provoke hyperammonemic crisis (Scaglia and Lee 2006). The diagnosis can be confirmed by laboratory findings of deficient arginase1 activity in erythrocytes and by molecular testing of the *ARG1* gene.

We describe two siblings with hyperargininemia and present the genetic investigation of their family, which yielded a novel *ARG1* mutation in an unusual site.

Material and Methods

Patient 1

A 9-year-old boy, the first child of consanguineous parents of Arabic origin, was referred to Schneider Children's Medical Center of Israel following hospitalization in another facility because of fever of one day's duration associated with a gradual deterioration in hepatocellular enzymes and coagulation functions. Past medical history revealed an uneventful pregnancy with birth weight 3.5 kg. The parents noticed protein aversion around the age of one year. Normal developmental milestones were achieved until age 2.5 years, when the patient began to exhibit a gradual loss of motor and verbal skills. By age 4 years, he was wheelchair-bound and had lost sphincter control. At age 5.5 years, he had two generalized tonic-clonic seizures (Table 1) with abnormal findings on electroencephalography. Magnetic resonance imaging scan was normal. Genetic consultation was noncontributory. The tentative diagnosis was cerebral palsy.

On physical examination, the patient was conscious. Severe mental retardation was noted; he could not speak and hardly responded to simple commands. No dysmorphism was noted. Vital signs were within normal range.

The liver was palpable 2 cm below the rib cage. Neurological examination revealed spasticity hyperreflexia and prominent contractures in the lower limbs (Table 1). Laboratory tests showed mild anemia and thrombocytopenia, with glutamic oxaloacetic transaminase 920 U/L, glutamic pyruvic transaminase 719 U/L, gamma-glutamyl transferase 74 U/L, lactic dehydrogenase 1257 U/L, international normalized ratio (INR) 1.8, creatine kinase 668 U/L, amylase 182 U/L, urea 25 mg/dL. Metabolic work-up was remarkable for high blood levels of arginine, high excretion of orotic acid in the urine, and mild to moderate hyperammonemia. Free and total carnitine levels were mildly low (Table 1). Blood citrulline, glutamine, and ornithine, blood gasses, lactate, pyruvate, and very-long-chain fatty acids were within normal limits.

The fever resolved within a few days, with gradual improvement in hepatocellular enzyme levels and INR. Arginase deficiency was presumed, and molecular investigation was performed.

Table 1 Main anthropometric, clinical, and metabolic characteristics of the two siblings with arginase deficiency

Characteristics	Patient 1	Patient 2
Sex	Male	Female
Age at onset/diagnosis	2–3/9 years	2–3/7 years
Anthropometric data[a]		
Birth weight (percentile)	3.5 kg (50)	3.0 kg(25)
Weight (percentile)	20.2 kg ($<$ 3)	19.8 kg ($<$ 3)
Head circumference (percentile)	51.5 cm (9.3)	51 cm (12)
Main areas of developmental deterioration and neurological manifestations	Mental, gross motor, language, sphincter control, epilepsy, hyperspasticity	Mental, gross motor, language, sphincter control, hyperspasticity
Blood ammonia (N 11–48 μmol/L)		
At onset	61 μmol/L	
At diagnosis	123 μmol/L	61 μmol/L
Arginine in plasma (N $<$ 140 μmol/L)	652 μmol/L	846 μmol/L
Serum free carnitine (N 25–45 μmol/L)	18.4 μmol/L	33.0 μmol/L
Serum total carnitine (N 30–50 μmol/L)	22.5 μmol/L	43.4 μmol/L
CSF arginine (N $<$ 35 μmol/L)	107 μmol/L	85 μmol/L
Urinary excretion		
Orotic acid (control $>$ 250 mmol/mol creatinine)	401 mmol/mol creatinine	Markedly elevated
Uracil (control 5-36 mmol/mol creatinine)	189 mmol/mol creatinine	Increased

[a] Height/length measurements could not be obtained because of the patients' severe contractures

Patient 2

A 7-year-old girl, the sister of patient 1, had a similar course of neurological deterioration and was therefore invited for investigation. Her anthropometric and clinical characteristics and metabolic profile are presented in Table 1.

Molecular Analysis

DNA samples of the patients, a third healthy sibling, and their parents were extracted from whole blood by standard methods. Linkage analysis was performed using microsatellite markers D6S262, D6S457, D6S413 (Fig. 1). Polymerase chain reaction (PCR) products were analyzed on an ABI3100 genetic analyzer with Genescan software (Applied Biosystems, Foster City, CA, USA). *ARG1* was sequenced by amplifying each exon and its flanking regions. PCR products were purified and sequenced with the BigDye terminator system on an ABI prism 3700 sequencer. As *ARG1* is constitutively expressed in granulocytes in addition to hepatocytes (Munder et al. 2005), RNA was extracted from freshly prepared leukocytes of the family members using the High Pure RNA isolation kit (Roche, Mannheim, Germany), and first-strand synthesis was performed with Superscript II (Invitrogen, Carlsbad, CA, USA). The primers used for RT-PCR and for the PCR of the identified mutation are listed in Table 2. PCR products for the mutation analysis were restricted with BspE1, which cuts the mutant allele. One hundred control alleles were tested to rule out polymorphism.

Results

Linkage analysis indicated that both affected children were homozygous for the same haplotype. The parents and healthy sib were carriers of this haplotype (Fig. 1).

DNA sequencing of each exon of the *ARG1* gene did not reveal any base change. However, RT-PCR products of fragment A of the patients (Fig. 2) showed a unique pattern: one band (c) with much weaker expression in the patients than in the mother and control despite its similar size, and two longer bands (a, b) expressed only in the patients. Sequencing of band c revealed a normal *ARG1* transcript with much lower expression than in the control. The upper band (a) contained a 125-bp insertion between exons 3 and 4, leading to the addition of 41 amino acids and a premature termination codon TGA at the sixth codon downstream to the inserted fragment. Band b was a mixture of bands a and c.

Blast analysis of the inserted fragment revealed that it originated from the third intron of the *ARG1* gene, at c.305 + 1314-_1438. The intronic sequence of the insertion was flanked by the consensus splice sites, AG and GT. An intronic mutation c.305 + 1323 t > c was found at the tenth nucleotide in the inserted fragment. The patients were homozygous for this mutation, and the parents and healthy sibling were heterozygous, as shown by PCR and restriction by BspE1 (Fig. 3). The mutation was not found in the DNA of 100 controls of Arabic origin.

Discussion

As in previous reports of argininemia (Prasad et al. 1997; Lee et al. 2011), our patients were initially thought to have cerebral palsy, which considerably delayed the correct diagnosis. Argininemia was presumed on the basis of the characteristic disease course and clinical findings (Crombez and Cederbaum 2005) combined with the typical biochemical results (Table 1).

Because of its prenatal diagnostic advantage, molecular analysis was preferred for confirmation over the red blood cell arginase activity assay.

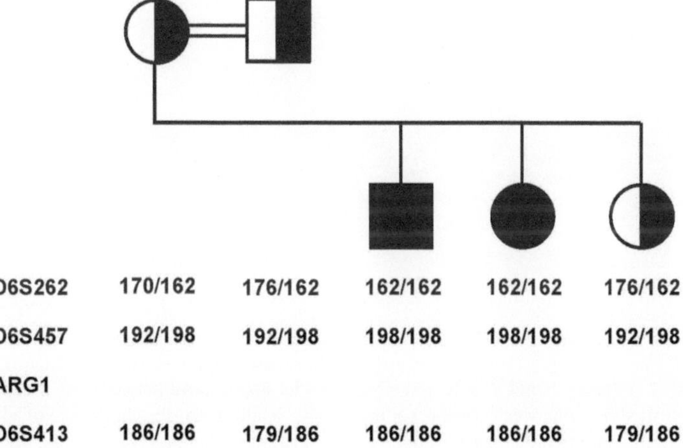

Fig. 1 Linkage analysis of the family. The two affected children are homozygous for the same haplotype, whereas the parents and the healthy sib are heterozygous for this haplotype

The novel mutation identified here is added to the approximately 20 known *ARG1* mutations (http:www. hgmd.cf.ac.uk) reported in patients of different ethnic origins. These include missense, nonsense, splicing, regulatory mutations deletions and insertions. Most mutations are private. However, there is one report of a possible ethnicity-associated mutation (R21X) in four unrelated Portuguese patients (Cardoso et al. 1999).

The reported *ARG1* mutations are located along the coding region of the gene (http:www.hgmd.cf.ac.uk). An association has been found between the type and site of these mutations and the structural and functional changes in the protein product (Vockley et al. 1996). Missense mutations within highly conserved regions that encodes for amino acids at the active site of arginase1 impairs enzyme function; substitution mutations outside these regions are thought to be silent with no clinical implications. By contrast, stop codon mutations and microdeletions may appear randomly along the gene and cause a truncated protein with consequently total loss of arginase1 activity (Vockley et al. 1996). We speculate that the chain-terminating mutation found in our patients generated a transcript harboring a truncated peptide leading to loss of enzyme activity.

The intronic t > c change, which was the only difference between the normal and the affected sequence, is an exon-splicing-enhancer mutation. It created a better binding site for different splicing factors that define AG and GT on both sides of the intronic inserted fragment as splice sites. Interestingly, the intronic t > c change was already noted when *ARG1* DNA was initially sequenced, but at that point, it was considered to be a clinically insignificant single nucleotide polymorphism.

The finding of a normal *ARG1* transcript (band c) in the patient's lane (Fig. 2), although in a much smaller amount than in his mother's and the control lanes is not surprising given the presence of the consensus donor and acceptor sites flanking all the *ARG1* exons . We believe that the

Table 2 Primers used for RT-PCR analysis of *ARG1* and for PCR of the identified mutation

	Primers
ARG1 open-reading frame	
Fragment A	ARG1AF 5′-TGACTGACTGGAGAGCTCAAG-3′ and ARG1AR 5′-GTCTGTCCACTTCAGTCAT TG-3′
Fragment B	ARG1BF 5′-TGTGTATATTGGCTTGAGAGA-3′ and ARG1BR 5′-TTGAATTTTACACCAAGAG GG-3′
Identified mutation	F-5′-CCGCAATACTTTTGCACCAAC-3′
	R-5′-CGTATGGCGTGTGTTCACTC-3′

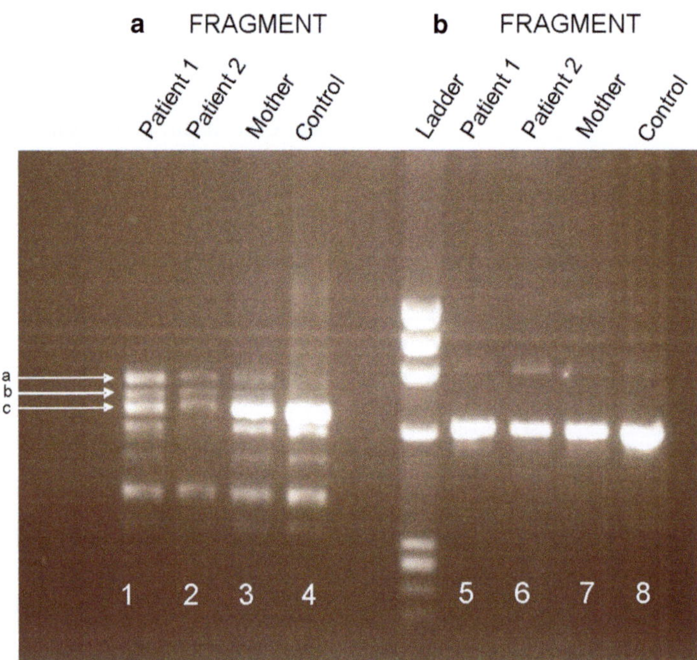

Fig. 2 RT-PCR of *ARG1*. Fragment A (lanes *1* and *2*, affected sibs; lane *3*, mother; lane *4*, normal control) shows one weak band (*c*) at the level of the control band in lanes *1* and *2* and two bands (*a, b*) that are absent in the control. Band *c* is characterized by a normal *ARG1* transcript, with much lower expression in the patients than the control. The upper band (*a*) contains a 125-bp insertion between exons 3 and 4, leading to the addition of 41 amino acids and a premature termination codon TGA at the sixth codon downstream. Band *b* is a mixture of bands *a* and *c*. Fragment B was similar in the affected sibs (lanes *5,6*), their mother (lane *7*), and the normal control (lane *8*)

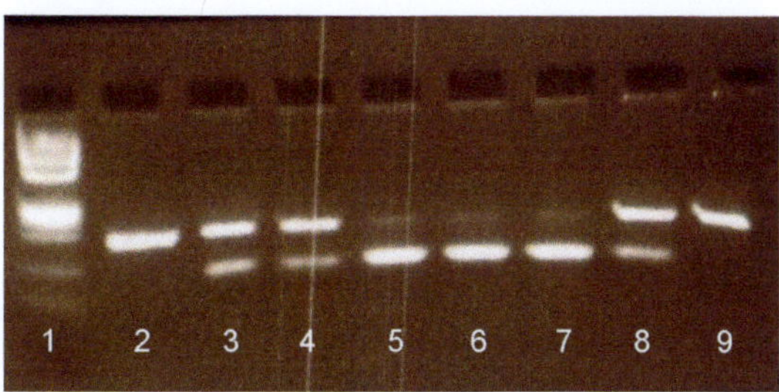

Fig. 3 Restriction analysis of the *ARG1* mutation c.305+1314_1438. The mutation introduces a BspE1 restriction site. Lane *1*, DNA ladder; lane *2*, unrestricted PCR product; lanes *3–9*, restricted PCR products: lane *3* – mother, lane *4* – father, lane *8* – healthy sib showing partial restriction indicating a heterozygous genotype, lanes *5* and *6* – duplicates of patient 1 demonstrating complete restriction for a homozygous genotype; lane *7* – patient 2, lane *9* – restricted control indicating absence of restriction site

reason this transcript does not prevent the disease derives from the homotrimeric composition of arginase1 (Brusilow and Horwich 2001), such that assembly of the normal protein with the mutant one does not maintain an active enzyme.

To our knowledge, this is the first reported exon-splicing-enhancer mutation in patients with hyperargininemia, and the second reported mutation in a family of Arab origin (Korman et al. 2004).

This report should alert clinicians to the possibility of missed diagnosis of hyperargininemia in children presenting with severe spastic paraparesis and mental retardation. It also highlights the significance of analyzing the expression pattern of a candidate gene if sequencing at the DNA level is unrevealing.

Take Home Message

A late diagnosis of hyperargininemia should be considered in pediatric patients with neurodevelopmental delay and regression in early childhood.

When cDNA sequencing is unrevealing, the expression pattern of the candidate gene should be analyzed for identification of potential novel mutations.

Details of Contributors

1. Dr. Haimi-Cohen collected and analyzed the clinical data, described the patients, and drafted and edited the manuscript.
2. Mrs. Bargal conducted and interpreted the molecular analysis and described the genetic findings in the manuscript.
3. Dr. Zeigler supervised the molecular analysis investigation and contributed significantly in interpreting the data and revising the manuscript.
4. Dr. Eidlitz-Markus helped in the collection and analysis of the clinical data and made a great contribution to the writing of the manuscript.
5. Mrs. Zuri contributed to all stages of the molecular analysis and the interpretation of its results and assisted in revising the manuscript.
6. Dr. Zeharia initiated the study. He contributed substantially to coordination of the clinical and molecular data of the study, to its design, and to the revision of the manuscript.

Guarantor of Study

Yishai Haimi-Cohen

Patient Consent

Not applicable

Competing Interests

None of the authors has a conflict of interest.

Funding

No funding was received for this study.

References

Braga AC, Vilarinho L, Ferreira E, Rocha H (1997) Hyperargininemia presenting as persistent neonatal jaundice and hepatic cirrhosis. J Pediatr Gastroenterol Nutr 24:218–221

Brusilow SW, Horwich AL (2001) Urea cycle enzymes. In: Sriver CR, Baudet AL, Sly WS, Valle D (eds) The metabolic bases of inherited diseases, 8th edn. McGraw-Hill, New York, pp 1909–1963

Cardoso ML, Martins E, Vasconcelos R, Vilarinho L, Rocha J (1999) Identification of a novel R21X mutation in the liver-type arginase gene (ARG1) in four Portuguese patients with argininemia. Hum Mutat 14:355–356

Crombez EA, Cederbaum SD (2005) Hyperargininemia due to liver arginase deficiency. Mol Genet Metab 84:243–251

De Deyn PP, Marescau B, Qureshi IA (1997) Hyperargininemia: A treatable inborn error of metabolism? In: De Deyn PP, Marescau B, Qureshi IA et al (eds) Guanidino compounds in biology and medicine, vol 2. John Libbey & Company Ltd, London, pp 53–69

Gomes Martins E, Santos Silva E, Vilarinho S, Saudubray JM, Vilarinho L (2011) Neonatal cholestasis: an uncommon presenta-

tion of hyperargininemia. J Inhert Metab Dis. [Epub ahead of print] doi: 10.1007/s10545-10-9263-7, Jan 13 Online

Korman SH, Gutman A, Stemmer E, Kay BS, Ben-Neriah Z, Zeigler M (2004) Prenatal diagnosis for arginase deficiency by second-trimester fetal erythrocyte arginase assay and first-trimester ARG1 mutation analysis. Prenat Diagn 24:857–860

Lee BH, Jin HH, Kim GH, Choi JH, Yoo HW (2011) Argininemia presenting with progressive spastic diplegia. Pediatr Neurol 44:218–220

Munder M, Mollinedo F, Calafat J et al (2005) Arginase 1 is constitutively expressed in human granulocytes and participates in fungicidal activity. Blood 105:2549–2556

Picker JD, Puga AC, Levy HL et al (2003) Arginase deficiency with lethal neonatal expression: evidence for the glutamine hypothesis of cerebral edema. J Pediatr 142:349–352

Prasad AN, Breen JC, Ampola MG, Rosman NP (1997) Argininemia: A treatable genetic cause of progressive spastic diplegia simulating cerebral palsy: case reports and literature review. J Child Neurol 12:301–309

Scaglia F, Lee B (2006) Clinical, biochemical and molecular spectrum of hyperargininemia due to arginase 1 deficiency. Am J Med Genet C Semin Med Genet 142C:113–120

Sparkes RS, Dizikes GJ, Klisak I et al (1986) The gene for human arginase (ARG1) is assigned to chromosome band 6q23. Am J Hum Genet 39:186–193

Vockley JG, Goodman BK, Tabor DE et al (1996) Loss of function mutations in conserved region of the human arginase 1 gene. Biochem Mol Med 59:44–51

JIMD Reports
DOI 10.1007/8904_2011_103

A Rare Galactosemia Complication: Vitreous Hemorrhage

Sahin Takci · Sibel Kadayifcilar · Turgay Coskun ·
Sule Yigit · Burcu Hismi

Received: 12 August 2011 / Revised: 20 September 2011 / Accepted: 12 October 2011 / Published online: 11 December 2011
© SSIEM and Springer-Verlag Berlin Heidelberg 2011

Abstract Galactosemia is a secondary glycosylation disorder characterized by galactose deficiency of glycoproteins and glycolipids. Abnormal glycosylation of coagulation factors and evidence of liver disease are associated with coagulopathy in galactosemic infants. We report a case of a neonate with galactosemia presenting with bilateral vitreous hemorrhage (VH). During the follow-up, hemorrhage in the right eye resolved; however, it persisted in the left eye. Vitrectomy was planned for the left eye. In addition to cataract, VH is another ophthalmic finding in galactosemia with serious sequelae such as amblyopia. Serious complications of coagulopathy in galactosemic infants can be prevented with early diagnosis and prompt treatment. Inclusion of galactosemia in the neonatal screening program offers an opportunity to prevent early severe symptoms.

Communicated by: Gerard T. Berry

Competing interests: None declared.

S. Takci (✉) · S. Yigit
Neonatology Unit, Hacettepe University Ihsan Dogramaci
Children's Hospital, 06100, Ankara, Turkey
e-mail: stakci@gmail.com

S. Kadayifcilar
Department of Ophthalmology, Hacettepe University Faculty
of Medicine, Ankara, Turkey

T. Coskun
Pediatric Metabolism, Hacettepe University Ihsan Dogramaci
Children's Hospital, Ankara, Turkey

B. Hismi
Department of Pediatric Metabolism, Hacettepe University Ihsan
Dogramaci Children's Hospital 06100, Ankara, Turkey

Abbreviations

GALE	UDP-galactose 4'epimerase
GALK	Galactokinase
GALT	Galactose 1-phosphate uridyltransferase
VH	Vitreous hemorrhage

Introduction

Galactose is metabolized by a series of sequential reactions collectively known as the Leloir pathway (Fig. 1). The three enzymes that catalyze these reactions are galactokinase (GALK), galactose 1-phosphate uridyltransferase (GALT), and UDP-galactose 4'epimerase (GALE). Galactosemia is caused by deficiency of any one of these enzymes, resulting in accumulation of galactose 1-phosphate and galactose in breast-fed or regular infant milk formula-fed infants (Fridovich-Keil 2006). The most common and clinically severe form is due to GALT deficiency, which is known as classic galactosemia, affecting about 1:30,000 to 60,000 live-births in the Caucasian population. It is an autosomal recessive disorder caused by a mutation in the *GALT* gene on the short arm of chromosome 9, with more than 150 disease-causing mutations (Bosch 2006; Item et al. 2002).

Compared to GALT deficiency, GALE and GALK deficiencies are rare. The incidence of GALK deficiency is less than 1:100,000 (Fridovich-Keil 2006).

Galactosemia usually presents as a life-threatening disease within the first weeks of life after ingestion of galactose. The initial clinical symptoms are generally nonspecific, like feeding difficulties, vomiting and diarrhea, lethargy, and hypotonia. Clotting abnormalities secondary to liver disease and predominantly *Escherichia*

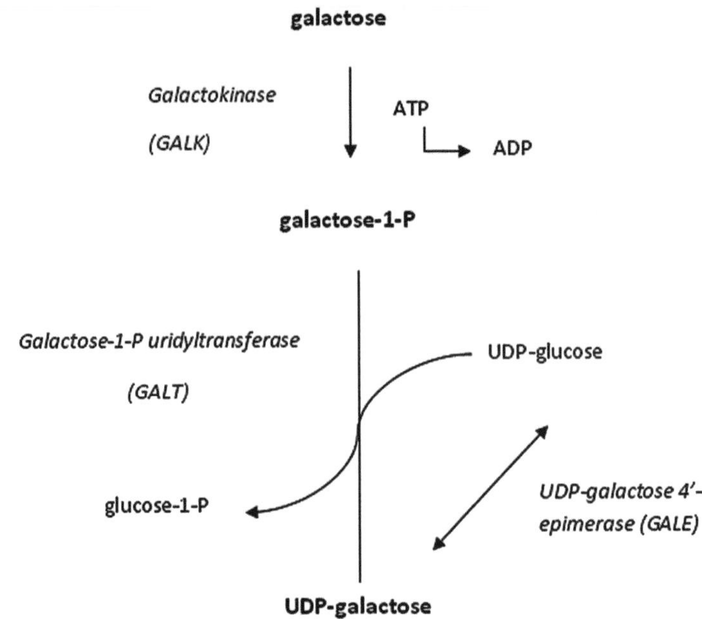

Fig. 1 The Leloir pathway of galactose metabolism

coli sepsis can lead to early death in untreated infants (Holtan et al. 2001). Affected infants are also at increased risk of delayed development, speech difficulties, and intellectual disability. Females with classic galactosemia may experience reproductive problems caused by ovarian failure (Bosch 2006). Diagnosis and treatment of galactosemia should be performed as early as possible in order to prevent neonatal death and to minimize the complications (Bosch 2006).

Cataract has been reported as the main ophthalmic finding of galactosemia (Burke et al. 1989). Apart from cataract, there are also reports on another ophthalmic complication, vitreous hemorrhage (VH), which is rarely observed (Levy et al. 1996). Here, we present bilateral VH, with spontaneous resolution in one eye, in an infant diagnosed with classic galactosemia.

Case Report

A 12-day-old female infant who was sent from a local hospital was admitted to the neonatal intensive care unit because of poor feeding, jaundice, and hepatomegaly. She was born at 40 weeks' gestation by spontaneous vaginal delivery to a 30-year-old gravida 6, para 3 mother, with a 3,500 g birth weight. The Apgar scores were 8 and 9 at 1 and 5 min, respectively. On the seventh day of life, she was hospitalized because of indirect hyperbilirubinemia and given intensive phototherapy for 2 days. Her parents were first cousins. The mother had two early abortions and a

female baby who had died at 7 days of age with indirect hyperbilirubinemia requiring exchange transfusion.

On admission, the infant was hypoactive with poor sucking. She had been receiving breast-milk exclusively. The vital signs were normal. Physical examination revealed insufficient weight gain (body weight 3,450 g), jaundice and palpable liver 4 cm below the right costal margin. No bleeding was noted from the mucosal membranes or venipuncture sites. Ophthalmologic examination revealed icteric scleras, reactive pupils, and the absence of bilateral red reflexes.

Hemoglobin was 14.9 g/dL, leukocytes 13.4×10^9/L, and platelets 47×10^9/L, and immature/total neutrophil ratio was 0.11 in total blood count on admission. The biochemical markers were as follows: glucose 56 mg/dL, serum aspartate aminotransferase 171 IU/L, alanine aminotransferase 96 IU/L, alkaline phosphatase 255 IU/L, gamma-glutamyl transferase 29.9 IU/L, total bilirubin/direct bilirubin: 19.4/13.7 mg/dL, albumin 2.7 g/dL, blood urea nitrogen 12.9 mg/dL, and creatinine 0.23 mg/dL. Activated partial thromboplastin time was 34.2 s, international normalized ratio was 0.95, and fibrinogen level was 280 mg/dL (219–403) on the 18th day of life. Serum C-reactive protein (5.4 mg/dL) and serum procalcitonin levels (8.1 ng/ml) were high. Urine culture was sterile; however, the blood culture was positive for *E. coli*. The treatment with ampicillin and gentamicin, which was started on the first day of admission, was continued. Follow-up blood culture was sterile after 5 days of the antibiotic treatment. Ursodeoxycholic acid was given for cholestasis. Arterial blood gases were normal. Plasma and

urinary amino acid analyses showed mild elevation of tyrosine and phenylalanine. The clinical suspicion of galactosemia was strengthened by positive reducing substance test in the urine and by the detection of galactose spot on urinary sugar chromatography. Low GALT activity in erythrocytes (2.7 U/dL; normal value: 21.2 U/dL) and a homozygous Q188R mutation on molecular genetic testing confirmed the diagnosis of classic galactosemia. Galactose-free formula was started on the 13th day of life and the infant's health status improved gradually. The patient was discharged on the 22nd day of life.

In addition to the metabolic workup, the patient was evaluated on the 13th day of life by a retina specialist, who performed a handheld slit-lamp biomicroscopy and fundus examination, revealing the absence of bilateral red reflexes, bilateral pupillary tunica vasculosa lentis, and bilateral VH. Repeated ophthalmic examination disclosed posterior synechia in the left eye and bilateral VH. There was no evidence of cataract, and intraocular pressures were normal. Ocular ultrasonography showed bilateral VH on the 30th day of life. The left globe was noted to be slightly smaller in size than the right. The last ocular examination revealed diffuse retinal pigment epithelial changes and resolving VH in the right eye on the 39th day of life. Visual evoked potentials for the right and left globe were normal. Vitrectomy was planned for the left eye. Follow-up visits were scheduled under the purview of a team composed of a metabolic physician, dietician, neonatologist, and pediatric ophthalmologist.

Discussion

Galactosemia is an autosomal recessive disorder caused by deficient or absent activities of any of the three enzymes involved in the galactose metabolic pathway. The predominant form is classic-type galactosemia, which is due to a diminished activity or absence of the GALT enzyme. This enzyme deficiency leads to accumulation of galactose and its metabolites, galactose-1 P and galactitol, in body tissues and fluids, and leads to a secondary glycosylation defect. Glycosylation is a post-translational modification that mediates the form and function of many proteins as coagulation factors. Hypoglycosylation of coagulation factors causes the abnormal coagulation function in galactosemia as well as the mechanism of congenital disorders of glycosylation type 1 in untreated patients (Sturiale et al. 2005).

Galactosemic infants often present with jaundice after initiation of lactose-containing formulas. On admission, indirect hyperbilirubinemia was prominent, and the patient required phototherapy. After the second week of life, direct bilirubin dominated. Woo et al. (2010) reported a case with early indirect hyperbilirubinemia requiring exchange transfusion diagnosed with galactosemia. In our case, the sibling who died (presumably due to undiagnosed galactosemia) also had indirect hyperbilirubinemia requiring exchange transfusion. It should be recognized that the early hyperbilirubinemia in untreated galactosemia is of the indirect type, and that direct hyperbilirubinemia – or less than indirect type – does not appear until about one week of age.

Galactose is converted to galactitol in cells and produces osmotic effects such as swelling of lens fibers, which may result in cataracts (Elsas 2000). Cataract is the main ocular sign associated with classic galactosemia. In an international survey study including 314 galactosemic patients, cataracts were reported in 30%. Only 8 patients diagnosed with galactosemic cataract required surgery; the remaining were mild or transient (Waggoner et al. 1990). It has been suggested in early reports that cataracts appeared with noncompliance with the diet; however, in the retrospective study of Widger et al. (2010), no direct relation between the degree of dietary compliance and cataract formation could be demonstrated. Current guidelines recommend a lifelong regular ophthalmic monitoring for cataract in galactosemic patients, but the period of ophthalmic examination is controversial.

VH is another ophthalmic complication of galactosemia, although its prevalence is unknown (Levy et al. 1996). The retinal vessel changes in galactose-fed dogs have been shown with a wide histopathologic spectrum in a number of animal studies. These retinal changes were associated with the occlusion of capillary beds and subsequent ischemia of the retina, including endothelial cell proliferation, soft exudates, intraretinal microvascular abnormalities and their subsequent degeneration, preretinal–intraretinal hemorrhages, and new vessel growth into the vitreous (Takahashi et al. 1992; Kador et al. 1995; Kobayashi et al. 1998). Galactosemic infants are susceptible to retinal changes probably leading to VH because of galactose and its metabolites.

Retinal hemorrhages are seen in nearly one-third of infants born by vaginal delivery. The incidence of retinal hemorrhage increases to 75% among infants delivered by vacuum extraction. Most retinal hemorrhages due to birth trauma usually resolve by 2 weeks of life (Emerson et al. 2001). Although retinal hemorrhage is frequent, VH is very rare in infants. It results from perinatal complications, disseminated intravascular coagulation, protein C deficiency, retinopathy of prematurity, Terson's syndrome, and shaken-baby syndrome (Ferrone and de Juan 1994).

VH as an ophthalmic complication in galactosemic infants is reported very rarely in the literature. Laumonier et al. (2005) reported a galactosemic infant with unilateral VH who was free from cataract. Levy et al. (1996) reported five galactosemic neonates with VH. Coagulopathy was demonstrated in two neonates, and one infant was noted to have small bruises and bleeding in venous access sites. They suggested that coagulopathy contributed to the retinal vessel fragility of the neonate, exacerbated by galactosemia. The obstetric history revealed no risk for VH in this case. The presence of intravascular coagulopathy could not be ascertained in this case, since the prothrombin and partial thromboplastin times were not determined until 5 days after the baby had been started on a galactose-free formula. VH in infants can have very serious sequelae, such as axial myopia, severe amblyopia, and tractional retinal detachment. These potential sequelae could occur within 5 weeks after dense hemorrhage. In infants, early surgical intervention with vitrectomy is recommended – after waiting 3 to 4 weeks for spontaneous clearance of a dense hemorrhage – because of the severity of the sequelae (Ferrone and de Juan 1994).

In addition to cataract, one has to consider VH in the ophthalmic examination of infants with galactosemia. Presence of coagulopathy may exacerbate retinal hemorrhage progressing to VH in these infants. Q188R is the most common mutation among Turkish classic galactosemia patients as well as all Caucasian populations (Coskun et al. 1995). This mutation is generally associated with a severe biochemical phenotype, with nearly undetectable residual erythrocyte GALT activity in homozygotes. Screening of galactosemia in Turkey will provide an opportunity to prevent early severe symptoms of coagulopathy with early dietary and supportive interventions (Tyfield et al. 1999).

Concise Summary

Cataract has been reported as the main ophthalmic finding of galactosemia. However, there are also reports on another ophthalmic complication, vitreous hemorrhage (VH), which is rarely observed. Here, we present bilateral VH, with spontaneous resolution in one eye, in an infant diagnosed with classic galactosemia.

Details of the Contributions of Individual Authors

Sahin Takci is the corresponding author and the guarantor for this paper.

Sibel Kadayifciler was responsible for the ophthalmic examination.

Turgay Coskun was responsible for the metabolic and genetic workup for galactosemia.

Sule Yigit followed the patient in the neonatal intensive care unit and helped in the writing of the manuscript.

Competing Interest

All authors declare that the answers to all questions on the *JIMD* competing interest questionnaire are "No."

Details of Funding

The authors confirm independence from the sponsors the content of the article has not been influenced by the sponsors.

Patient Consent Form

A patient consent statement about the patient proof that informed consent was obtained.

References

Bosch AM (2006) Classical galactosaemia revisited. J Inherit Metab Dis 29:516–525

Burke JP, O'Keefe M, Bowell R, Naughten ER (1989) Ophthalmic findings in classical galactosemia-a screened population. J Pediatr Ophthalmol Strabismus 26:165–168

Coskun T, Erkul E, Seyrantepe V, Ozgüç M, Tokatli A, Ozalp I (1995) Mutational analysis of Turkish galactosaemia patients. J Inherit Metab Dis 18:368–369

Elsas LJ (2000) Galactosemia. http://www.ncbi.nlm.nih.gov/book-shelf/br.fcgi?book=gene&part=galactosemia

Emerson MV, Pieramici DJ, Stoessel KM, Berreen JP, Gariano RF (2001) Incidence and rate of disappearance of retinal hemorrhage in newborns. Ophthalmology 108:36–39

Ferrone PJ, de Juan E Jr (1994) Vitreous hemorrhage in infants. Arch Ophthalmol 112:1185–1189

Fridovich-Keil JL (2006) Galactosemia: the good, the bad, and the unknown. J Cell Physiol 209:701–705

Holtan JB, Walter JH, Tyfield LA (2001) Galactosemia. In: The metabolic and molecular bases of inherited disease, 8th edn. The McGraw-Hill Companies, Newyork

Item C, Hagerty BP, Mühl A, Greber-Platzer S, Stöckler-Ipsiroglu S, Strobl W (2002) Mutations at the galactose-1-p-uridyltransferase gene in infants with a positive galactosemia newborn screening test. Pediatr Res 51:511–516

Kador PF, Takahashi Y, Wyman M, Ferris F (1995) Diabetes like proliferative retinal changes in galactose-fed dogs. Arch Ophthalmol 113:352–354

Kobayashi T, Kubo E, Takahashi Y, Kasahara T, Yonezawa H, Akagi Y (1998) Retinal vessel changes in galactose-fed dogs. Arch Ophthalmol 116:785–789

Laumonier E, Labalette P, Morisot C, Mouriaux F, Dobbelaere D, Rouland JF (2005) Vitreous hemorrhage in a neonate with galactosemia. A case report. J Fr Ophtalmol 28:490–496

Levy HL, Brown AE, Williams SE, de Juan E Jr (1996) Vitreous hemorrhage as an ophthalmic complication of galactosemia. J Pediatr 129:922–925

Sturiale L, Barone R, Fiumara A et al (2005) Hypoglycosylation with increased fucosylation and branching of serum transferrin N-glycans in untreated galactosemia. Glycobiology 15: 1268–1276

Takahashi Y, Wyman M, Ferris F, Kador PF (1992) Diabetes like preproliferative retinal changes in galactose-fed dogs. Arch Ophthalmol 110:1295–1302

Tyfield L, Reichardt J, Fridovich-Keil J et al (1999) Classical galactosemia and mutations at the galactose-1-phosphate uridyl transferase (GALT) gene. Hum Mutat 13:417–430

Waggoner DD, Buist NR, Donnell GN (1990) Long-term prognosis in galactosaemia: results of a survey of 350 cases. J Inherit Metab Dis 13:802–818

Widger J, O'Toole J, Geoghegan O, O'Keefe M, Manning R (2010) Diet and visually significant cataracts in galactosaemia: is regular follow up necessary? J Inherit Metab Dis 33:129–132

Woo HC, Phornphutkul C, Laptook AR (2010) Early and severe indirect hyperbilirubinemia as a manifestation of galactosemia. J Perinatol 30:295–297

JIMD Reports
DOI 10.1007/8904_2011_104

CASE REPORT

Neonatal Cholestasis as Initial Manifestation of Type 2 Gaucher Disease: A Continuum in the Spectrum of Early Onset Gaucher Disease

Abdallah F. Elias · Maria Ronningen Johnson ·
John K. Boitnott · David Valle

Received: 27 August 2011 /Revised: 10 October 2011 /Accepted: 12 October 2011 /Published online: 11 December 2011
© SSIEM and Springer-Verlag Berlin Heidelberg 2011

Abstract Gaucher disease type 2 [OMIM #230800] is a rare lysosomal storage disorder with usual onset between 3 and 6 months of age leading to progressive neurodegeneration and death within the first 2 years of life. Rarely it may lack the characteristic symptom-free period and initially manifest prenatally or in the neonatal period. The early course of neonatal onset classic type 2 variants is not well known, and reports of early histological changes in the liver of type 2 Gaucher disease patients are scarce. We describe a patient who presented in the immediate postnatal period with cholestasis without hepatomegaly associated with hepatocellular giant-cell transformation on liver biopsy, thrombocytopenia, and failure to thrive. This was initially thought to represent neonatal giant-cell hepatitis and the correct diagnosis was not made until the age of 6 months. Hepatocellular giant transformation has not been described in the classic acute neuronopathic form of GD. However, it has been reported in congenital GD with nonimmune hydrops and neonatal hepatitis, an example of perinatal lethal Gaucher disease (PLGD), which sometimes is regarded as an entity separate from GD type 2. Our case illustrates that neonatal cholestasis may be part of a spectrum of manifestations which spans a continuum between the PLGD and classic type 2 GD. Giant cells are a nonspecific finding but may reflect the

presence of a systemic inflammatory process that recently has been implicated in the brain stem degeneration associated with acute neuronopathic GD.

Introduction

Three major phenotypes of Gaucher disease (OMIM #230800) have been described based on the absence (type 1) or presence (types 2 and 3) of central nervous system involvement. Type 2, also known as acute neuronopathic form, is rare, with an estimated frequency of 1/150,000 (Sidransky 1997). It typically presents between 3 and 6 months after birth with rapidly progressing neurological degeneration and visceral signs resulting in death prior to age 2 years (Grabowski and Beutler 2006). The early course of classic type 2 variants with neonatal onset is not well known. Here, we report the case of neonatal cholestasis associated with nonspecific giant-cell transformation on liver biopsy as initial manifestation of acute neuronopathic Gaucher disease.

Case History

The male infant was born at an outside hospital (OSH) to nonconsanguineous, African–American parents at 39 weeks gestation via C-section for breech presentation after an otherwise unremarkable pregnancy. His birth weight was 3,500 g. Postnatally, he developed jaundice with a conjugated bilirubin level of 8 mg/dl [136.8 μmol/l] on the first day of life reaching maximum levels of conjugated bilirubin of 15.5 mg/dl [265 μmol/l] and total bilirubin of 17.8 mg/dl [304.38 μmol/l]. Liver transaminases showed maximal elevation for aspartate aminotransferase (AST) to 670 U/l and for alanine aminotransferase (ALT) to 330 U/l. Thrombocytope-

Communicated by: Maurizio Scarpa.

Competing interests: None declared.

A.F. Elias (✉) · M.R. Johnson · D. Valle
Department of Pediatrics, Johns Hopkins Hospital, McKusick-Nathans Institute of Genetic Medicine, 600 N. Wolfe St., Blalock 1008, BaltimoreMD 21287-4922, USA
e-mail: aelias2@jhmi.edu

J.K. Boitnott
Department of Pathology, Johns Hopkins School of Medicine, Baltimore, MD 21231, USA

nia with platelets of 45,000/mm^3 was treated with platelet transfusion within the first week of life while anemia was monitored reaching a hemoglobin level of 9.8 mg/dl within the first month. The report of a liver biopsy performed at 3½ weeks of life noted multinucleated hepatocytes (giant cells), a minimal intralobular inflammatory infiltrate and mild peripheral bile duct proliferation with cholestasis. Hepatic architecture was preserved and hepatocellular necrosis absent. Genetic evaluation during the newborn period at the outside hospital (OSH) was inconclusive yielding essentially normal results for plasma alpha-1 antitrypsin, plasma amino acids, urine organic acids, total and free carnitine, acylcarnitine profile, bile acids, GALT, very long chain fatty acids, and karyotype. At 6 months of age, the child was seen at the OSH for persistent failure to thrive. His weight was 4.79 kg (50th percentile for 6 weeks old), length 59 cm (50th percentile for 2 months old), head circumference 39.8 cm (50th percentile for 3 months old). A comparative genomic hybridization array, and a brain MRI were unremarkable.

Two weeks later, the infant was admitted to our service after a prolonged apnea episode. He had significant developmental delay. Neurological examination showed axial and appendicular hypertonicity and retroflexion of the neck with opisthotonus. Bulbar signs included ophthalmoplegia as well as dysphagia. The recurrent apnea episodes were not associated with stridor and suspected to be a manifestation of brain stem dysfunction in the absence of seizure activity in the EEG. His abdomen was protuberant with severe hepatosplenomegaly. A large, nonincarcerated umbilical hernia was present. Initial laboratory studies were significant for anemia, mild thrombocytopenia, and elevated liver transaminases, but the hyperbilirubinemia had completely resolved. The combination of central nervous system involvement, psychomotor retardation, hepatosplenomegaly and thrombocytopenia suggested the possibility of a lysosomal storage disease. Furthermore, the pattern of neurological impairment with markedly increased tonus and prominent bulbar signs pointed to the neuronopathic form of Gaucher disease as a primary differential diagnosis. A bone marrow aspirate did not contain the typical lipid-laden macrophages with fibrillary structures (Gaucher cells). However, a repeat liver biopsy now showed numerous enlarged histiocytes with cytological features typical of Gaucher cells and the giant-cell transformation seen in the earlier biopsy specimen was no longer apparent (Fig. 1a). The β-glucosylceramidase activity measured in the patient's leukocytes on two different blood specimens was 0.6 and 0.3 nmol/h/mg, respectively (reference range: 4–9 nmol/h/mg). The results indicated a profoundly reduced β-glucosylceramidase activity and confirmed the clinically suspected diagnosis of Gaucher disease. Sequence analysis of GBA, the gene encoding β-glucosylceramidase revealed a compound heterozygous genotype: one missense allele p.D409H

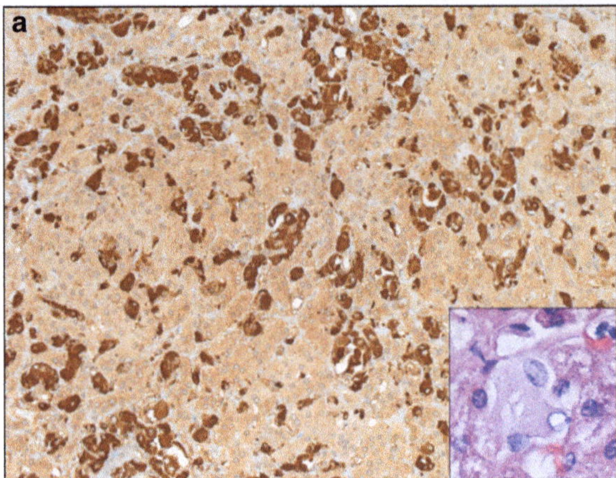
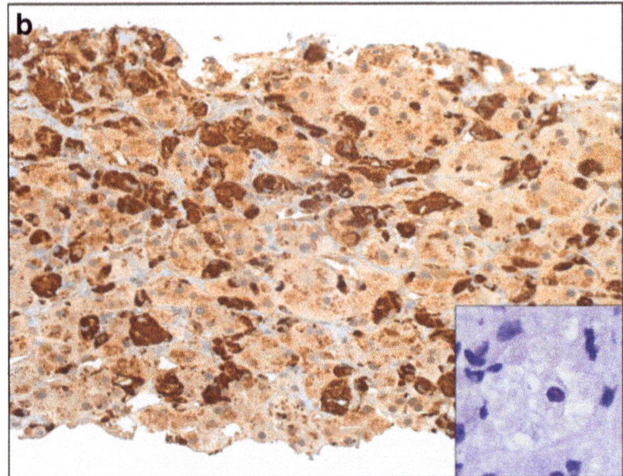

Fig. 1 CD68 stains with inserts of representative hematoxylin and eosin (H&E) stained histiocytes of liver biopsy specimen from the patient. (**a**), Liver biopsy specimen at 6 months of age when child initially presented to our service. Distortion of normal hepatocellular plate pattern was noted and numerous enlarged histiocytes with cytologic features typical of Gaucher cells are apparent. (**b**), Liver biopsy at 3½ weeks performed at the outside hospital. Paraffin blocks were obtained and stained in our pathology department. The pathology report from the OSH noted the presence of multinucleated hepatocytes (giant cells) and mild cholestasis with general preservation of the hepatic architecture. We used an antibody to CD68 to stain hepatic macrophages. This revealed more numerous and widespread enlarged sinusoidal histiocytes than are typically present in neonatal cholestasis, but these did not have the typical appearance of Gaucher cells (H&E insert)

(c.1342G>C) in exon 9; the other allele harbored three variants in *cis*, p.L444P (c.1448T>C), p.A456P (c.1483G>C), and p.V460V (1497G>C). In the following months, there was a progressive decline of the child's neurological status with the development of marasmus. Death occurred at the age of 14 months after a short course of pneumonia.

Case Discussion

We describe a rare course of neonatal Gaucher disease in a patient who presented in the immediate postnatal period with cholestasis without hepatomegaly associated with nonspecific liver histology, thrombocytopenia, and severe failure to thrive. This was initially thought to represent neonatal giant-cell hepatitis. The correct diagnosis was not made until the age of 6 months, at what time he displayed the full spectrum of neurologic and visceral involvement associated with type 2 GD.

Rarely GD type 2 may lack the characteristic symptom-free period and initially manifest prenatally or in the neonatal period. A distinct phenotype termed perinatal lethal Gaucher disease (PLGD) has emerged (Mignot et al. 2003). In PLGD, clinical signs are present at birth and include arthrogryposis, ichthyosis, and facial dysmorphy. Nonimmune hydrops fetalis or *in utero* demise may be prominent features (Orvisky et al. 2002). Some authors regard PLGD as an entity separate from Gaucher disease type 2 based on the presence or absence of clinical signs at birth as well as the clinical spectrum of manifestation (Mignot et al. 2006).

In contrast to PLGD, there is a scarcity of information about the course of neonatal onset classic type 2 variants. As our case illustrates, atypical presentation during the early neonatal period can make an early diagnosis difficult. Predominant features at birth in this patient were direct hyperbilirubinemia, elevation of transaminases and thrombocytopenia in the absence of clinically detectable hepatosplenomegaly. Cholestasis was transient and associated with nonspecific histological changes on liver biopsy (Jevon and Dimmick 1998), which included hepatocellular giant cells.

We identified two previous case reports of early onset Gaucher disease with neonatal cholestasis and hepatomegaly and subsequent development of neurological signs consistent with the acute neuronopathic form (type 2) (Barbier et al. 2002; Schwartz et al. 2009). Details of a liver biopsy performed within the first month of life were available only for one report and, in contrast to our case, did not reveal giant-cell transformation (Barbier et al. 2002).

Three cases of neonatal cholestasis associated with severe courses of perinatal Gaucher disease reminiscent of PLGD were previously reported. The patients had facial dysmorphism, but no neurologic involvement. (Ben Turkia et al. 2009). One case presented with severe thrombocytopenia requiring platelet transfusions, but only mild direct hyperbilirubinemia peaking at 5.4 mg/dl and a normal neurological exam at 5 months of age (Roth et al. 2005).

These cases illustrate that neonatal cholestasis may be part of a spectrum of manifestations, which spans a continuum between the distinct phenotypes of PLGD and classic type 2 Gaucher disease. Detailed descriptions of early histological changes in the liver of type 2 GD patients are scarce (Abrahamov et al. 1995). Hepatocellular giant transformation observed in the initial liver biopsy from our case has not been described in the classic acute

neuronopathic form of Gaucher disease, but has been reported in congenital Gaucher disease with nonimmune hydrops and neonatal hepatitis (Spear et al. 2007) and is a known feature of other lysosomal storage diseases such as Niemann–Pick Type C (Jevon and Dimmick 1998). It is a frequent component of neonatal hepatitis occurring in all cholestatic conditions of infancy and as such is a nonspecific finding consistent with some degree of inflammation (Crawford 2006). Review of the initial liver pathology using an antibody to CD68 expressed in macrophages revealed more numerous and widespread enlarged sinusoidal histiocytes than are typically present in neonatal cholestasis, but these did not have the typical appearance of Gaucher cells (Fig. 1b). These findings suggest that inflammatory processes may be important in early pathogenesis and that the clinical phenotype is due to an effect of macrophage storage beyond the physical presence of the Gaucher cells (Cox 2001). Various proinflammatory cytokines have been implicated in the pathogenesis of Gaucher disease including IL-1 (Gery et al. 1981), TNF-alpha, IL-6, and IL-10 (Michelakakis et al. 1996; Allen et al. 1997). In the Gaucher mouse model, accumulation of glucocerebroside appears to mediate brain inflammation via proinflammatory cytokines (Hong et al. 2006). This raises the possibility that the initial nonspecific clinical and histological features could be manifestations of a systemic inflammatory response (SIR) and as such harbinger of an inflammatory process ultimately resulting in the brain stem degeneration associated with the acute neuronopathic course of GD.

Almost 300 mutations and several polymorphisms have been reported in the GBA gene, but the specific genotypes do not correlate well with the phenotype (Hruska et al. 2008; Sidransky 2004). Most mutations are rare and rarely found in the homozygous form. A recombinant allele containing the three point mutations as identified in our patient — two that introduce the amino acid substitutions L444P and A456P, and one silent mutation, V460V, in exon 10 – was first reported in 1990 and named RecNciI (Eyal 1990; Latham 1990). Recombinant alleles are essentially null alleles and despite being found in all three types of GD are associated with more severe disease manifestations (Tayebi et al. 2003; Cormand et al. 1998).

In summary, neonatal cholestasis presenting with nonspecific histological findings on liver biopsy can be a rare initial manifestation of acute neuronopathic GD. In the absence of typical neurological or visceral findings, a high index of suspicion is needed to make the diagnosis by measuring β-glucosylceramidase activity in lymphocytes or fibroblasts. Lack of Gaucher cells in liver or bone marrow specimen does not exclude GD. Molecular genetic analysis of the genotype may be of prognostic value and help with clinicopathological correlation in certain cases.

References

Abrahamov A, Elstein D, Gross-Tsur V, Farber B, Glaser Y, Hadas-Halpern I et al (1995) Gaucher's disease variant characterised by progressive calcification of heart valves and unique genotype. Lancet 346:1000–1003

Allen MJ, Myer BJ, Khokher AM, Rushton N, Cox TM (1997) Proinflammatory cytokines and the pathogenesis of Gaucher's disease: increased release of interleukin-6 and interleukin-10. QJM 90:19–25

Barbier C, Devisme L, Dobbelaere D, Noizet O, Nelken B, Gottrand F (2002) Neonatal cholestasis and infantile Gaucher disease: a case report. Acta Paediatr 91:1399–1401

Ben Turkia H, Tebib N, Azzouz H, Abdelmoula MS, Ben Chehida A, Caillaud C et al (2009) Phenotypic continuum of type 2 Gaucher's disease: an intermediate phenotype between perinatal-lethal and classic type 2 Gaucher's disease. J Perinatol 29:170–172

Cormand B, Harboe TL, Gort L, Campoy C, Blanco M, Chamoles N et al (1998) Mutation analysis of Gaucher disease patients from Argentina: high prevalence of the RecNciI mutation. Am J Med Genet 80:343–351

Cox TM (2001) Gaucher disease: understanding the molecular pathogenesis of sphingolipidoses. J Inherit Metab Dis 24 (Suppl 2):106–121, discussion 187–108

Crawford JM (2006) Basic mechanisms in hepatopathology. In: Burt AD, Portmann BC, Ferrell LD (eds) MacSween's pathology of the liver, 5th edn. Churchill Livingstone Elsevier, Philadelphia

Gery I, Zigler JS Jr, Brady RO, Barranger JA (1981) Selective effects of glucocerebroside (Gaucher's storage material) on macrophage cultures. J Clin Invest 68:1182–1189

Grabowski GA, Beutler E (2006) Chapter 146: Gaucher disease. In: Valle D, Beaudet AL, Vogelstein B, Kinzler KW, Antonarakis SE (eds) The online metabolic and molecular bases of inherited disease. McGraw-Hill, New York, Chapter 146

Hong YB, Kim EY, Jung SC (2006) Upregulation of proinflammatory cytokines in the fetal brain of the Gaucher mouse. J Korean Med Sci 21:733–738

Hruska KS, LaMarca ME, Scott CR, Sidransky E (2008) Gaucher disease: mutation and polymorphism spectrum in the glucocerebrosidase gene (GBA). Hum Mutat 29:567–583

Jevon GP, Dimmick JE (1998) Histopathologic approach to metabolic liver disease: Part 1. Pediatr Dev Pathol 1:179–199

Michelakakis H, Spanou C, Kondyli A, Dimitriou E, Van Weely S, Hollak CE et al (1996) Plasma tumor necrosis factor-a (TNF-a) levels in Gaucher disease. Biochim Biophys Acta 1317:219–222

Mignot C, Gelot A, Bessieres B, Daffos F, Voyer M, Menez F et al (2003) Perinatal-lethal Gaucher disease. Am J Med Genet A 120A:338–344

Mignot C, Doummar D, Maire I, De Villemeur TB (2006) Type 2 Gaucher disease: 15 new cases and review of the literature. Brain Dev 28:39–48

Orvisky E, Park JK, LaMarca ME, Ginns EI, Martin BM, Tayebi N et al (2002) Glucosylsphingosine accumulation in tissues from patients with Gaucher disease: correlation with phenotype and genotype. Mol Genet Metab 76:262–270

Roth P, Sklower Brooks S, Potaznik D, Cooma R, Sahdev S (2005) Neonatal Gaucher disease presenting as persistent thrombocytopenia. J Perinatol 25:356–358

Schwartz IV, Krug B, Picon PD (2009) Commentaire de l'article "Cholestase neonatale revelatrice d'un phenotype intermediaire d'une maladie de Gaucher type 2". Arch Pediatr 16:1190–1191

Sidransky E (1997) New perspectives in type 2 Gaucher disease. Adv Pediatr 44:73–107

Sidransky E (2004) Gaucher disease: complexity in a "simple" disorder. Mol Genet Metab 83:6–15

Spear GS, Beutler E, Hungs M (2007) Congenital Gaucher disease with nonimmune hydrops/erythroblastosis, infantile arterial calcification, and neonatal hepatitis/fibrosis. Clinicopathologic report with enzymatic and genetic analysis. Fetal Pediatr Pathol 26:153–168

Tayebi N, Stubblefield BK, Park JK, Orvisky E, Walker JM, LaMarca ME et al (2003) Reciprocal and nonreciprocal recombination at the glucocerebrosidase gene region: implications for complexity in Gaucher disease. Am J Hum Genet 72:519–534

JIMD Reports
DOI 10.1007/8904_2011_105

Clinical and Biochemical Profiles of Maple Syrup Urine Disease in Malaysian Children

**Z Md. Yunus · DP Abg Kamaludin · M Mamat ·
YS Choy · LH Ngu**

Received: 25 August 2011 /Revised: 10 October 2011 /Accepted: 13 October 2011 /Published online: 11 December 2011
© SSIEM and Springer-Verlag Berlin Heidelberg 2011

Abstract *Introduction*: Maple Syrup Urine Disease (MSUD) is an autosomal recessive disorder caused by defects in the branched-chain α-ketoacid dehydrogenase complex resulting in accumulation of branched-chain amino acids (BCAAs) and corresponding branched-chain ketoacids (BCKAs) in tissues and plasma, which are neurotoxic. Early diagnosis and subsequent nutritional modification management can reduce the morbidity and mortality. Prior to 1990s, the diagnosis of MSUD and other inborn errors of metabolism (IEM) in Malaysia were merely based on clinical suspicion and qualitative one-dimensional thin layer chromatography technique. We have successfully established specific laboratory diagnostic techniques to diagnose MSUD and other IEM. We described here our experience in performing high-risk screening for IEM in Malaysia from 1999 to 2006. We analysed the clinical and biochemical profiles of 25 patients with MSUD.

Methods: A total of 12,728 plasma and urine samples from patients suspected of having IEM were received from physicians all over Malaysia. Plasma amino acids quantitation using fully automated amino acid analyzer and identification of urinary organic acids using Gas Chromatography Mass Spectrometry (GCMS). Patients' clinical information were obtained from the request forms and case records

Results: Twenty-five patients were diagnosed MSUD. Nineteen patients (76%) were affected by classical MSUD, whereas six patients had non-classical MSUD. Delayed diagnosis was common among our case series, and 80% of patients had survived with treatment with mild-to-moderate learning difficulties.

Conclusion: Our findings suggested that MSUD is not uncommon in Malaysia especially among the Malay and early laboratory diagnosis is crucial.

Communicated by: Daniela Karall.

Competing interests: None declared.

Z. Md. Yunus (✉) · D.P. Abg Kamaludin · M. Mamat
Biochemistry Unit, Specialised Diagnostic Centre, Institute for
Medical Research, Jalan Pahang, 50588 Kuala Lumpur, Malaysia
e-mail: zabedah@imr.gov.my

Y.S. Choy
Prince Court Medical Centre, Kuala Lumpur, Malaysia

Y.S. Choy
Paediatric Institute, Hospital Kuala Lumpur, Kuala Lumpur,
Malaysia

L.H. Ngu
Medical Genetic Department, Hospital Kuala Lumpur, Jalan
Pahang, 50588 Kuala Lumpur, Malaysia

Introduction

Maple Syrup Urine Disease (MSUD) is an autosomal recessive disorder caused by defects in the mitochondrial branched-chain α-ketoacid dehydrogenase complex (BCKAD). The BCKAD is a large complex with four subunits (E1α, E1β, E2 and E3) and is necessary for decarboxylation of branched-chain ketoacids (BCKAs), which is the second step in the degradation pathway of branched-chain amino acids (BCAAs) leucine, isoleucine and valine. Mutations in both alleles encoding any subunit can result in the accumulation of BCAAs, alloisoleucine and their corresponding BCKAs measured in blood, urine and cerebrospinal fluid (Fig. 1) (Chuang et al. 2001; Wappner and Gibson 2002).

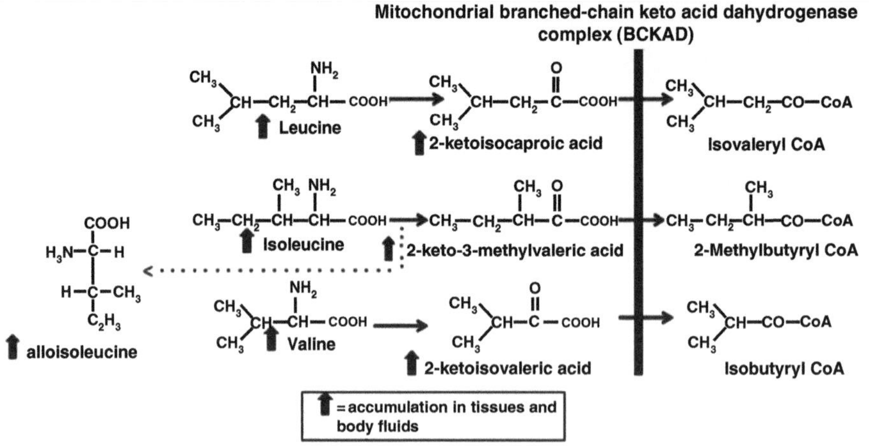

Fig. 1 Metabolic pathway of branched-chain amino acids degradation. In MSUD, the degradation of the essential branched-chain amino acids leucine, valine and isoleucine and their derived 2-ketoacids is impaired because of deficiency of BCKAD. Alloisoleucine which is a by-product of isoleucine transaminatione significantly elevated

The clinical manifestation of MSUD depends on the severity of BCKAD deficiency. Patients with classic MSUD typically have enzyme activity less than 2%. They present with poor feeding, lethargy, irritability, maple syrup-like or burnt sugar smell and ketonuria in first week of life. Without treatment, they develop progressive neurological deterioration due to cerebral edema, culminating in coma, central respiratory failure and death. In patients with non-classical MSUD, residual enzyme activity varies from 2% to 30%, resulting in delayed clinical presentation to infancy or childhood as feeding problems, poor growth, developmental delay and behavioural problems. Non-classical patients may also experience severe metabolic intoxication and encephalopathy if sufficiently stressed by infectious illness, dehydration or prolonged fasting (Chuang et al. 2001). Early diagnosis and subsequent dietary modification by restricting BCAAs can rescue patients with MSUD from the devastating neurological complications (Chuang et al. 2001; Wappner and Gibson 2002).

The worldwide frequency of MSUD based on the data of newborn screening programme from 26.8 million newborns is approximately 1 in 185,000 live births. In countries where consanguineous marriage is common like Saudi Arabia, Turkey, Spain and India, the frequency is higher (Chuang et al. 2001). In South East Asia, the frequency is not known (Lee et al. 2008; Pangkanon et al. 2008). In Malaysia, MSUD was previously thought to be a rare disease.

Prior to 1990s, the diagnosis of MSUD and other inborn errors of metabolism (IEM) in Malaysia were merely based on clinical suspicion and detection of increased BCAAs and other amino acids in urine by one-dimensional thin layer chromatography (TLC). This approach was undesirable because TLC is only a qualitative method, laborious, time-consuming and interpretation of the results are operator-dependent and often inconclusive due to interference from a lot of substances in patient's urine such as drugs and bilirubin. Most importantly, alloisoleucine, the diagnostic marker for MSUD, cannot be detected using TLC. In late 1990s, we have successfully established an ion-exchange chromatography method for quantitative analysis of amino acids, as well as a qualitative analysis of urinary organic acid using Gas Chromatograpy Mass Spectrometry (GCMS) method. From 1999 to 2006, we have performed amino acids and/or organic acids analyses in 12,728 patients' samples using this new diagnostic approach. We described here our experience in performing high-risk screening for IEM in Malaysia. We analysed the clinical and biochemical profiles of 25 patients with MSUD, which was one of the commonest IEM diagnosed during this period.

Materials and Methods

Specimen

Physicians who wished to request IEM screening for their patients filled in a special request form in which patient's clinical signs and symptoms and basic laboratory results are documented (Fig. 2). Blood from suspected patients was collected in heparin centrifuged immediately and the plasma was separated. Random urine was collected in a sterile bottle and immediately frozen after collection. Both specimens were transported to the laboratory in ice to prevent degradation of amino acids and organic acids at higher temperature. Plasma and urine from normal control were also analysed for reference.

Equipments

Plasma amino acids were analysed by ion-exchange chromatography technique using a fully automated amino acids analyzer, Biochrom 30+ which used EZChrom Elite V2B software for quantitation, manufactured by Biochrom Ltd., Cambridge, UK. Organic acids were analysed using GCMS system and Chemstation software manufactured by Agilent Technologies Inc, California, USA. All standards, solvents and chemicals required for the analyses were purchased either from Sigma Aldrich, Missouri, USA or Merck KGaA Darmstadt. Germany. Physiological amino acids kits were purchased from Biochrom Ltd, Cambridge UK. All organic solvents and other reagents used were analytical grade.

Sample Processing

For the plasma amino acid analysis, 100 μL of plasma in eppendorf tube was deproteinised with 100 μL of 10% sulphosalicylic acids. The tube was capped and vortexed for a few seconds and kept for 1 h at 4°C. After 1 h, the sample was centrifuged for 5 min. The supernatant was filtered through 0.2 μm membrane filter to remove any remaining particulate materials prior to analysis. About 20 μL of this supernatant was then loaded by autosampler into a column of cation-exchange resin. Buffers of varying pH and ionic strength were then pumped through the column to separate the various amino acids. The column temperature was accurately controlled and varied as necessary to produce the required separation of all 40

IEM REQUEST FORM

<u>**IMPORTANT NOTICE:**</u> To ensure correct, reliable result and interpretation given, the following must be followed:

1. Please fill up the entire form.
2. At least 1ml plasma and 5ml urine are needed. Heparinised plasma is prefered.
3. Seperate plasma from RBC immediately. Haemolysed samples will be rejected.
4. ALL processed samples (plasma and urine) must be frozen immediately and transport in <u>DRY ICE</u> to IMR.

Name:_______________________ Age:______ Sex: M / F / U Race: M / C / I / O_______

RN:____________M I/C:________________ Hospital:___________ Wad:__________

House Address:____ Tel:______________

1. Symptoms/signs of current illness:

Fever		Poor sucking/feeding		
Pallor		Respiratory problem		
Jaundice		Difficulty in breathing		
Hypothermia		Mental retardation		
Hypotonia/floppy		Developmental delay		
Cyanosed		Failure to thrive		
Lethargy		Feeding intolerance		
Easily irritable		Septicaemic-like illness		
Seizures or h/o seizures		Headache		
Drowsy		Smelly urine		
Coma		Colored urine		
Abnormal behaviour		Skin lessions		
Frequent vomiting		Eye lessions		

Other symptoms/signs:____

2. <u>Feeding history:</u> Type of milk: Breast/ Formula/ Mixed/ Solid diet:________________

2. <u>Family history:</u> Consanguinity: Yes/ No. If Yes please specify:________________

Occurrence of in	Stillbirth	neonatal death	neonatal seizures	metabolic disease
Siblings				
Maternal side				
Paternal side				

4. <u>Physical Examination:</u>

Respiratory distress		Hyperreflexia		
Dysmorphic features		Nystagmus		
Hypothermia		Optical atrophy		
Cardiomyopathy		Ptossis		
Drowsy		Abnormal odour		
Coma		Abnormal hair		
Opisthotonus		Hepatomegaly		
Dystonia		Splenomegaly		
Choreoathetoid movement		Eczema/ Othor rashes		
Hyptonia		Others (specify)		

Fig. 2 Continued

<u>5. Treatment given (specimen should be taken before any form of treatment given or stop for 2-3 days)</u>

Drug therapy: Antibiotic: No/Yes________________ Anticonvulsant: No/Yes________________

        Steroid: No/Yes________________ Other drug:________________________

Fluid infusion: Saline / Dextrose / Mannitol / Parenteral feeding / Others:________________

<u>6. Lab Result (before treatment is given)</u>

LFT: ALT:______________mmol/l   Blood Glucose:__________mmol/l

     AST:______________mmol/l   Blood Ammonia__________

     ALP:______________mmol/ l   Blood Lactate:______________pyruvate:________

Blood Gases: Normal / Met. Acidosis / Met.alkalosis / Resp.acidosis / Resp.alkalosis

Anion Gap:____________________Other relevant test (specify)________________

CTscan/MRI:____

Urine Analysis: pH__________Ketones: Pos/ Neg Reducing Sugar: Post/Neg

Provisional Diagnosis:________________________________

<u>7. Test required:</u> (Please tick appropriate test/s required)

1.	Plasma amino acid	
2.	CSF amino acid	
3.	Urine Orotic acid	
4.	Urine Organic Acid	
5.	Urine GAG / MPS (DMB & HRE)	
6.	Plasma Total and Free Carnitine	
7.	Plasma Total homocysteine	
8.	Blood spot: Total Galactose	
9.	Blood spot: GALT / GIPUT	
10.	Blood spot: AA & Acylcarnitines (IEM screening)	
11.	Plasma VLCFA and Phytanic acid	
12.	Urine Succinylacetone	
13.	Urine S-sulphocysteine	
14.	Urine Oligosaccharides	
15.	Urine Delta ALA	
16.	Urine Phorphyrin & Phorphobilinogen	
17.	Urine Cystine	
18.	Urine Homocystine	
19.	Others (Please specify):	

Date specimen collected: ____________________

Collected by: ____________________

Date specimen send: ____________________

Paediatrician In-Charge ____________________

Sign. And chop:

Fig. 2 Special request form for investigation of Inborn Errors of Metabolism

amino acids. The column eluent was mixed with the ninhydrin reagent, and passed through the high temperature reaction coil. In the reaction coil, ninhydrin reacted with the amino acids present in the eluate to form coloured compounds. The amount of coloured compound produced is directly proportional to the quantity of amino acid present in the eluate and detected using UV detector at two wavelengths, 570 nm and 440 nm, depending on the type of the amino acids concern (Biochrom 30 Instruction manual. Issue 6. Sect. 1, pp 1.2–1.3). A running time of 3 h was required for one complete run of amino acids profile. Mixed standards containing basic, acidic, and neutral amino acids were used to quantitate the amount of amino acids in the patient's sample.

Random urine for organic acids analysis was subjected to organic solvent extraction and derivatisation using method described by Gates et al. (1978), Tanaka et al. (1980), and Majors (1998) with minor modification in our laboratory (Pertiwi et al. 1999). In this method, urine sample was oximated with hydroxyl-amine. An internal standard, pentadecanoic acid, was added and later the organic acids content in the urine was extracted using ethyl-acetate and diethyl-ether. It was later derivatised with BSTFA [N,O-bis(trimethylsily)trifluoroacetamide] and injected into GCMS.

Clinical Information

Patients' clinical information was obtained from the request forms and case records. The following data were obtained: sex, ethnic group, consanguinity, age of onset of symptoms, clinical manifestations at initial diagnosis and outcome.

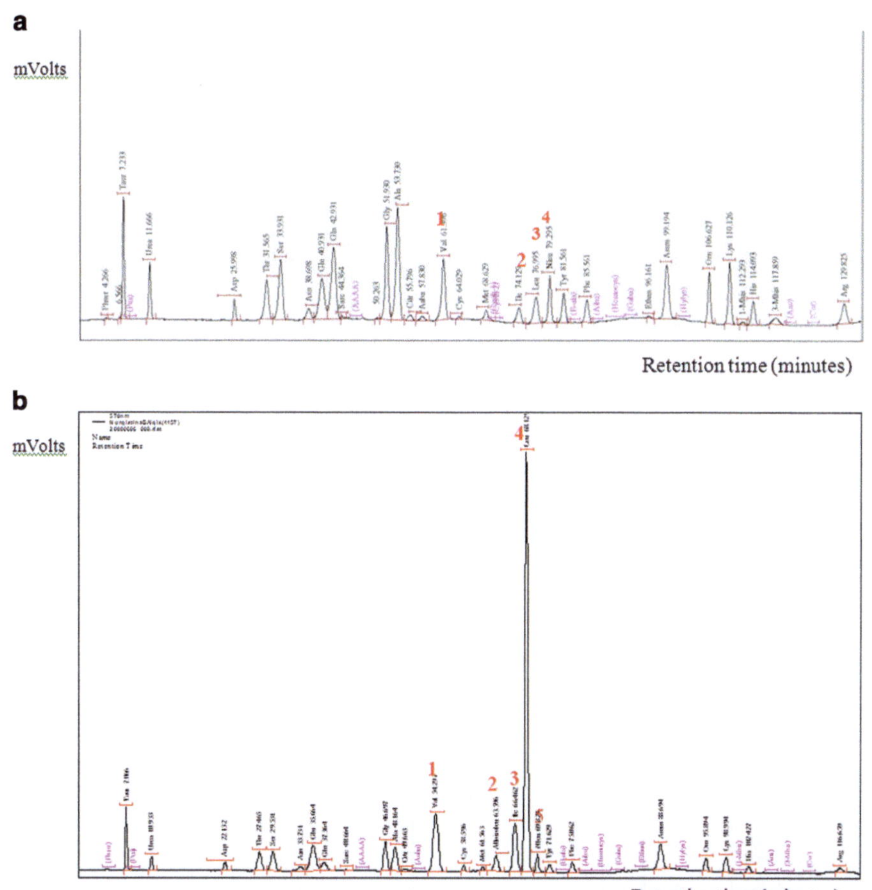

Fig. 3 Plasma amino acid chromatograms of (a) a normal control and (b) a patient with MSUD. (a) Plasma amino acid chromatogram from a normal control. Peak *1*, valine; *2*, isoleucine; *3*, leucine; *4*, internal standard (norleucine). Alloisoleucine is not detected. (b) Plasma amino acid chromatogram from a patient with MSUD showing markedly increased leucine. Plasma isoleucine and valine were also elevated. Alloisoleucine was detected. Peak *1*, valine; *2*, alloisoleucine; *3*, isoleucine; *4*, leucine; *5*, internal standard (norleucine)

Results

From 12,728 patients' samples that were analysed, approximately 2% of the samples yielded abnormal results. Twenty-five patients (13 males and 12 females) were diagnosed with MSUD based on elevated BCAAs with the presence of alloisoleucine in plasma and/or increased BCKAs in urine organic acids analysis. Figure 3a, b illustrates the plasma amino acids chromatogram from a normal control and one of the patients with MSUD, respectively. Their corresponding urine organic acids chromatogram is shown in Fig. 4a, b.

Among these 25 patients, 19 patients (76%) fall under the classical MSUD group, whereas six patients have non-classical MSUD based on the age of onset of initial symptoms. Parental consanguinity was found in one-third of the cases. Twenty-one cases (84%) were Malay, two cases were Chinese, and two cases were Bidayuh (an indigenous subpopulation in Sarawak). Case #7 and #8 are identical twin.

In the classical MSUD group, all patients developed initial symptoms within first week. Clinical manifestations were mostly non-specific and included poor feeding, breathing abnormalities, seizures and coma. Burnt sugar smell was present in nearly all patients by retrospective questioning of their parents. In 15 patients, the diagnosis was made within 1 month; whereas four other patients had their diagnosis made after neonatal period, ranged from 3 months to 12 months. Four patients died during the initial presentation. Learning difficulties were common among the survivors. The clinical and biochemical details of this group are summarised in Table 1. Plasma amino acids were done for case #5 and #9, but the results were not traceable. Clinical presentations showed that these two patients belong to the classical type.

Six patients have non-classical MSUD and presented with chronic neurological symptoms included developmental delay and seizure. Four of them have acute metabolic intoxication episode precipitated by febrile illness and one of them succumbed during the acute episode. Table 2 shows summary of clinical and biochemical details of these six patients.

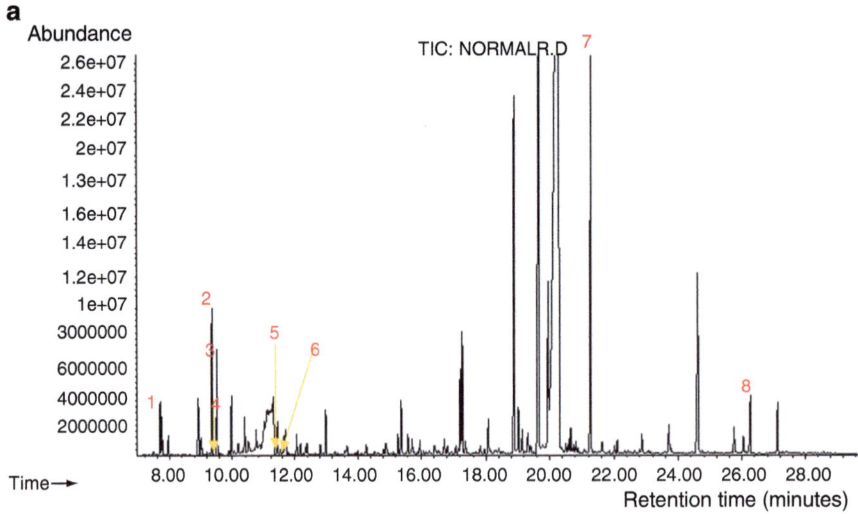

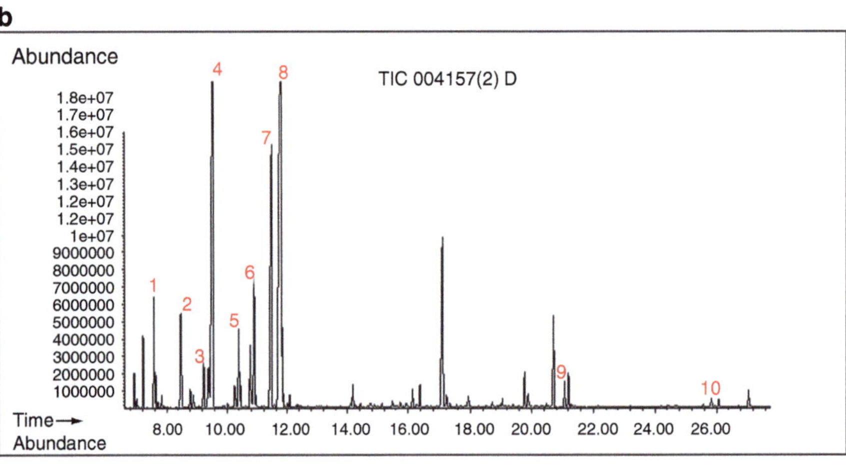

Fig. 4 Urinary organic acids chromatogram of (**a**) a normal control (**b**) a patient with MSUD. (**a**) Urine organic acids chromatogram from a normal control. Peak *1*, lactate; *2*, pyruvate; *3*, 3-hydroxybutyrate; *4*, 2-hydroxyisovalerate; *5*, 2-keto-3-methylvalerate; *6*, 2-keto-isocaproate; *7* and *8* are internal standard and external standard respectively. (**b**) Urine organic acids chromatogram from a patient with MSUD showing branched-chain ketoacids were excreted in large quantity in urine. Peak *1*, lactate; *2* and *8*, 2-keto-isocaproate (*2 peaks*); *3*, 3-OH butyrate; *4*, 2-hydroxyisovalerate; *5*, 2-ketoisovalerate; *6*, 2-hydroxy-isocaproate; *7*, 2-keto-3-methylvalerate; *9* and *10*, internal standard and external standard, respectively

Discussion

Definitive laboratory diagnosis is crucial in confirming clinical suspicion of MSUD since its clinical presentation is non-specific and can mimic common conditions such as infections and other IEMs. With the availability of quantitative plasma amino acid analysis and qualitative urine organic acids analysis in our centre, we have started to see an increasing number of children diagnosed to have MSUD in Malaysia. In our case series, MSUD is found to occur predominantly in a Malay population (84%). Most cases are classical type, comparable to a report from the Philippines where also the majority is in the classical group (Lee et al. 2008). Another noteworthy finding was that parental consanguinity was found in 1/3 of cases. Further genetic study is worthwhile consideration to find out the possible explanation about the higher prevalence of MSUD among Malay.

Delayed diagnosis is common in our case series. We believe that the delayed diagnosis is contributed by the lack of awareness among physicians. IEM was only being suspected when patient did not respond to treatment for common diseases such as infections. Very few physicians are able to identify clinically the special burnt sugar smell associated with MSUD, although this was present in almost all patients in our case series. The problem of delayed diagnosis is compounded by lack of diagnostic facilities. Our fully automated amino acids analyzer can only analyse maximum eight samples in a day. The number of samples

Table 1 Clinical and biochemical characteristics of patients with classical MSUD

Case	Age at diagnosis	Ethnic	Sex	Consanguinity	Clinical presentation	Leucine	Isoleucine	Valine	Elevated BCKA	Outcome
#1	13d	Malay	M	No	Coma, poor feeding, difficulty in breathing, FTT, sepsis	3,675	244	406	NA	Moderate MR
#2	31d	Malay	M	No	Poor feeding, FTT, sepsis	4,770	504	1,532	NA	Severe MR
#3	12m	Malay	M	Yes	Seizure at day 11, spastic	5,093	1,053	1,870	NA	Severe MR
#4	28d	Malay	F	No	Seizure, vomiting, difficulty in breathing, FTT	3,218	646	806	Yes	Severe MR
#5	3m	Malay	F	No	Sepsis, poor feeding, developmental delay	NA	NA	NA	Yes	Died
#6	1m	Bidayuh	F	No	Seizure, coma, poor feeding	3,390	646	806	Yes	Died during crises in neonatal period
#7	14d	Malay	M	Yes	Seizure, poor feeding, FTT, difficulty breathing	4,126	794	1,185	Yes	Well, normal IQ
#8	15d	Malay	F	No	Seizure, poor feeding, difficulty breathing	3,908	534	798	Yes	Died during crises in neonatal period
#9	26d	Malay	M	No	Floppy, poor feeding	NA	NA	NA	Yes	Moderate MR
#10	19d	Chinese	F	No	Difficulty breathing, sepsis	3,476	497	691	Yes	Moderate MR
#11	9d	Bidayuh	F	No	Poor feeding, difficulty breathing	3,572	419	386	Yes	Severe MR
#12	11d	Malay	F	No	Seizure, poor feeding	1,115	187	486	Yes	NA
#13	11d	Malay	M	No	Hypotonia, poor feeding, difficulty breathing, sepsis	3,305	405	577	Yes	NA
#14	12m	Malay	F	Yes	Seizure, coma, developmental delay, FTT	1,812	349	718	Yes	moderate MR
#15	22d	Malay	M	No	Poor feeding, coma, floppy, developmental delay	902	237	622	Yes	mild MR
#16	1m	Malay	M	Yes	Poor feeding, coma, developmental delay, FTT	2,473	728	1,128	Yes	mild MR
#17	1m	Malay	M	Yes	Poor feeding, coma, developmental delay, FTT	1,287	191	457	Yes	mild MR
#18	4.7m	Malay	F	Yes	Seizure, poor feeding, developmental delay, FTT	2,463	455	765	Yes	died
#19	6m	Malay	M	No	Poor feeding, seizure, developmental delay	NA	NA	NA	Yes	NA

d day, *m* month, *F* female, *M* male, *NA* not available, *FTT* failure to thrive, *MR* mental retardation

Table 2 Clinical and biochemical characteristics of patients with non-classical MSUD

Case	Age at diagnosis	Ethnic	Sex	Consanguinity	Presentation	Leu	Ile	Val	Elevated BCKA	Outcome
#20	5y	Malay	F	Yes	Developmental delay	921	284	386	NA	Mild MR
#21	3y	Malay	F	Yes	Mental retardation, developmental delay	611	152	264	Yes	Died during one of the crisis episodes
#22	1.3y	Malay	F	No	Seizure, developmental delay	2,658	753	1,690	NA	Moderate MR
#23	10y	Malay	M	No	Coma, mental retardation, developmental delay	1,713	337	999	Yes	Mild MR
#24	7y	Malay	M	No	Lethargy, drowsy	1,307	493	906	Yes	NA
#25	1y	Chinese	M	No	Drowsy, developmental delay. Not walking and talking	776	231	551	Yes	Walking and talking soon after treatment started. Now attending normal school, mild learning difficulty

y year, *NA* not available, *FTT* failure to thrive, *MR* mental retardation

we received far exceeded the capacity of our laboratory. The use of a special request form has made it easier for us to identify cases that require urgent attention. Cases with high index of suspicion of IEM such as clinical history of acute encephalopathy after a symptom-free period, positive family history; and abnormal routine biochemistry results such as persistent metabolic acidosis, recurrent hypoglycemia, hyperammonemia and ketosis were prioritised for immediate analysis. The awareness and understanding of physicians about MSUD should be heightened through continuous medical education and publicity.

Other researchers have suggested that leucine and its metabolite, α-ketoisocaproic acids, may be the main neurotoxic metabolites in MSUD since increased plasma concentration of these metabolites are associated with the appearance of neurological symptoms (Chuang et al. 2001; Funchal et al. 2007). Plasma leucine is invariably elevated in symptomatic MSUD patients, typically >1,000 μmol/L when there are acute neurological symptoms. The mean level of plasma leucine at the time of diagnosis in our case series was 3,037 μmol/L and 1,331 μmol/L, respectively, for classical and non-classical group. Plasma isoleucine and valine are also typically elevated, but may be normal or reduced. Plasma alloisoleucine, which is a distinctive metabolite present in all forms of MSUD, was elevated in all patients tested in our case series.

MSUD is a treatable genetic disease. Renal replacement methods such as hemodialysis or hemofiltration can achieve rapid corrections of BCAAs and BCKAs during the acute phase of MSUD crisis. Current long-term therapy for MSUD includes lifelong maintenance of a low BCAAs diet and regular monitoring of plasma BCAAs level. Aggressive treatment of intercurrent illness is necessary to prevent a catabolic state that can rapidly lead to release of BCAAs from muscle protein. With good treatment, patients with MSUD can reach adulthood with normal intelligence. However, mild-to-moderate learning difficulties are common (Chuang et al. 2001; Wappner and Gibson 2002).

Early diagnosis by the expanded newborn screening programme using Tandem Mass Spectrometry (TMS) in the developed countries followed by early intervention during the presymptomatic or early symptomatic period have been shown to improve the outcome of patients with MSUD (Simon et al. 2006; Yoon et al. 2003; Heldt et al. 2005). This should probably be the way forward for our country. If the screening procedure is properly applied (blood sampling at day 2–3 of life, sending the filter paper with dry blood spot without any delay and by overnight mail to a newborn screening laboratory, biochemical testing at the same day – notification of the attending physician and the parents of an abnormal test result) the tentative diagnosis of MSUD should be possible by days 3–5 and thus sufficiently early for adequate intervention.

Conclusion

MSUD is probably not uncommon in Malaysia especially among the Malay subpopulation. The majority of patients has classical MSUD and presented with acute neurological symptoms within first week of life. Although we are able to diagnose and manage MSUD, we recognise that the clinical outcome remains to be optimised. We should aim towards earlier diagnosis through improving accessibility to diagnostic facilities, increasing awareness among physicians and general public and establishing a newborn screening programme.

Acknowledgements We thank the Director General of Health Malaysia for permission to publish this paper. We would also like to express our gratitude to all Biochemistry staff for their technical assistance and to Dr Zakiah Ismail for initiating the project. Our special thanks to Dr Shahnaz Murad, Director of IMR, Dr. Rohani Md Yasin, Head of Specialized Diagnostic Centre and Dr Mohd Helmi Ismail for critical reading of the manuscript and valuable comments.

References

Chuang DT, Shih VE (2001) Maple syrup urine disease (branched-chain ketoaciduria). In: Scriver CR Beaudet AL, Sly WS, Valle D (eds), Childs B, Kinzler KW, Vogelstein B (assoc eds). The Metabolic and Molecular Basis of Inerited Disease, 8th edn. Mc Graw-Hill, New York, pp 1971–2006

Funchal C, Tramontina F, Quincozes dos Santos A, Frage de Souza D, Goncalves CA, Pessoa-Pureur R, Wajner M (2007) Effect of the branched-chain alpha-keto acids accumulating in maple syrup urine disease on S100B release from glial cells. J Neurol Sci 260 (1–2):87–94

Gates SC, Sweely CC, Krivit W, Dewitt D, Blaisdell BE (1978) Automated metabolic profiling of organic acids in human urine, II. Analysis of urine samples from healthy adults, sick children and children with neuroblastoma. Clin Chem 24(10): 1680–1689

Heldt K, Schwahn B, Marquardt I, Grotzke M, Wendel U (2005) Diagnosis of MSUD by newborn screening allows early intervention without extraneous detoxification. Mol Genet Metab 84(4):313–316

Lee JY, Chiong MA, Estrada SC, Cutingco-De la Paz EM, Silao CLT, Padilla CD (2008) Maple syrup urine disease (MSUD)-Clinical profile of 47 Filipino patients. JIMD Short report #135 Online

Majors RE (1998) Liquid extraction technique for sample preparation. LCGC Asia Pacific 1(2):10–15

Pangkanon S, Charoensiriwatana W, Sangtawesin V (2008) Maple syrup urine disease in Thai infants. J Med Assoc Thai 91(Suppl 3):S41–S44

Pertiwi AKD, Suwaji S, Zabedah MY, Zakiah I (1999) Screening for organic acidurias: method development using capillary gas chromatography with flame ionisation detector (FID) identification. Proceedings for 9th Annual Scientific Meeting of Malaysian Association of Clinical Biochemists

Simon E, Fingerhut R, Baumkötter J, Konstantopoulou V, Ratschmann R, Wendel U (2006) Maple syrup urine disease: favourable effect of early diagnosis by newborn screening on the neonatal course of the disease. J Inherit Metab Dis 29(4):532–537

Tanaka K, Hine DG, West-Dull A, Lynn TB (1980) Gas chromatographic method for analysis of urinary organic acids.1. Retention indices of 155 metabolically important compounds. Clin Chem 16(13):1839–1846

Wappner RS, Gibson KM (2002) Disorders of leucine metabolism. Physician's guide to the laboratory diagnosis of metabolic disease, 2nd edn. Springer, Berlin, pp 165–189

Yoon HR, Lee KR, Kim H, Kang S, Ha Y, Lee DH (2003) Tandem mass spectrometric analysis for disorders in amino, organic and fatty acid metabolism: two year experience in South Korea. Southeast Asian J Trop Med Public Health 34(Suppl 3): 115–120

JIMD Reports
DOI 10.1007/8904_2011_106

Severe Infusion Reactions to Fabry Enzyme Replacement Therapy: Rechallenge After Tracheostomy

K. Nicholls · K. Bleasel · G. Becker

Received: 02 June 2011 / Revised: 27 July 2011 / Accepted: 17 October 2011 / Published online: 11 December 2011
© SSIEM and Springer-Verlag Berlin Heidelberg 2011

Abstract A 34-year-old male patient with Fabry disease (OMIM 301500) commenced enzyme replacement therapy (ERT) with Agalsidase alfa, with positive clinical response. Infusion reactions, initially mild and easily managed, commenced during his 13th infusion, and continued over the next 3 years. Severity of reactions subsequently increased despite very slow infusion, extended prophylactic medication and attempted desensitisation, requiring regular intensive care unit (ICU) admissions. Facial oedema and flushing, throat tightness, headache and joint pain typically occurred 4–36 h after completion of most infusions, responding rapidly to subcutaneous adrenaline. Low titre specific IgG seroconversion was noted at 12 months, with subsequent reversion to negative after 5 years, despite persistence of infusion reactions. Specific IgE and skin testing was negative. Trial of ERT product switch to Agalsidase-beta resulted in no improvement in reactions. At 5 years, ERT was ceased in the face of recurrent ICU readmissions. In the face of progressive clinical deterioration, he underwent tracheostomy to allow recommencement of ERT. Two years later, he has clinically improved on regular attenuated dose Agalsidase-beta, administered by slow infusion in a local hospital setting.

Case Report

Diagnosed in early childhood following proband identification, the patient endured severe disabling neuropathic pain and diarrhoea during childhood and adolescence, truncating his educational and employment opportunities. Serum and WBC α-Galactosidase (GLA) levels were measured as 0.03 nmol/min/mg (normal range 0.4–2.0), and genotype was identified as G128E. At the time of enzyme replacement therapy (ERT) commencement at age 34, his BMI was 20.1 kg/m^2. His baseline pain was partially controlled on phenytoin, but frequent exacerbations occurred with infections and weather changes, sometimes requiring hospitalisation and narcotics. Diarrhoea was intractable, depression was significant, exercise capacity was limited to getting through his workday, and proteinuria had reached 800 mg/day, although Cr-EDTA GFR, ECG and echocardiogram were normal.

The first 12 infusions of ERT (Agalsidase alfa 0.2 mg/kg/fortnight over 40 min) were uneventful. The 13th infusion, delayed for 3 months for logistic reasons, was complicated by facial flushing and subjective throat tightness, without change in temperature or blood pressure. Symptoms resolved rapidly with cessation of infusion, intravenous hydrocortisone and promethazine.

All subsequent infusions were given under prophylaxis with hydrocortisone, combined variably with antihistamine (promethazine or certrizine), oral prednisolone and paracetamol. Over the next 3 years, reactions were occasional and mild, but accelerated in severity and frequency throughout

Communicated by: Robert J Desnick.

Competing interests: None declared.

K. Nicholls (✉) · G. Becker
Department of Nephrology, The Royal Melbourne Hospital, Parkville, VIC 3050, Australia
e-mail: kathy.nicholls@mh.org.au

K. Bleasel
Department of Immunology, The Royal Melbourne Hospital, Parkville, VIC 3050, Australia

K. Nicholls · K. Bleasel · G. Becker
The University of Melbourne, Parkville, VIC 3050, Australia

the fourth and fifth years of ERT, against a variety of attempted interventions. These included dose reduction, extended infusion times, and pre-treatment with various dose combinations of steroids, antihistamine, paracetamol, pethidine, transhexamic acid and danazol. Several attempts at desensitisation failed. Reactions requiring adrenaline treatment were induced with doses of Agalsidase alfa below 50 µg.

Reactions typically comprised facial oedema, throat tightness, headache, with variable flushing and joint pain. They were commonly delayed 4–36 h after completion of the infusion, and responded to the addition of adrenaline to steroid and antihistamine.

Baseline testing for IgG against GLA was negative, but seroconversion was first noted at low titre (maximum 1:100) at 12 months, with subsequent titres stable until 5 years after ERT initiation, when IgG reverted to negative. Specific testing revealed no evidence of IgE antibodies, mast cell or complement activation, or C1 esterase inhibitor deficiency, either during or after reactions. Skin tests were negative. During and after reactions, BP, oxygen saturation remained normal. Serial spirometry performed before, during, after and independent of ERT showed similar results – airway obstruction was minimal and bronchodilator response was negligible. Inhaled bronchodilator at the time of reactions did not help.

Specialist fibre-optic airway examination revealed no obvious structural cause for the symptoms experienced during ERT, but a narrow oropharynx with (Mallampati score IV) and large nasal polyps. Polysomnography revealed severe obstructive sleep apnoea (apnoea: hypopnoea index of 45 events per hour increasing to 93 events per hour in REM associated with significant oxygen desaturation ($PaO_2 = 76\%$) and sleep fragmentation).

Other than Fabry disease, no other risk factors for obstructive sleep apnoea were present. BMI had decreased slightly to 19.2 kg/m^2. The patient was very keen to proceed with ERT. His response had been subjectively positive, and ongoing daily diaries documented fewer and less severe pain exacerbations, less diarrhoea, and increased active life participation compared with his pre-ERT status. GFR was stable, and proteinuria reduced on renin–angiotensin blockade to 300–500 mg/day; however, the frequency and severity of reactions were clearly of major concern.

From Infusion 120 (5 years), ERT was changed to half-dose (0.5 mg/kg) Agalsidase-beta, which was initially well tolerated but induced a severe reaction on the second administration.

The dose was then further reduced from 35 to 10 mg, and infused over 10 h. When two consecutive treatments (Infusions 128 and 129) induced severe delayed reactions

requiring ICU readmission 24–48 h post-infusion, ERT was ceased, against the patient's expressed wishes, but in the interests of safety. Over the 2 years after cessation of ERT, the patient's pain worsened despite increased prophylactic therapy with maximal doses of multiple agents under advice from a pain management team, exercise tolerance decreased, sweating ceased and depression recurred. His ability to discern warmth in his distal lower limbs regressed to pre-ERT levels. Prior to ERT commencement, he was consistently unable to detect temperature sensation below the knees, but this had improved after 2 years of ERT, to the level of the mid metatarsals. Depression became increasingly significant, but medication was ineffective. Plasma Gb3 had increased from 3.9 nmol/ml at 5 years post ERT initiation, at which time he had been receiving only very low Agalsidase alpha dose (<0.1 mg/kg/fortnight) for the previous 6 months, to 7.3. In consultation with a multidisciplinary team, he elected to undergo tracheostomy to increase the safety margin to resume ERT, as well as to alleviate his sleep apnoea and upper airway obstruction.

Subsequently, 2.5 years after completely ceasing ERT and 7.5 years after his first ERT infusion, he underwent Agalsidase-beta rechallenge within a high-dependency setting, at a fortnightly dose of 10 mg over 12 h, under prophylaxis with hydrocortisone, certrizine and alprazolam). Mild to moderate reactions repeatedly occurred 1–12 h after completion of infusions. Premedication was changed from hydrocortisone to dexamethasone (to exclude the unlikely scenario that allergy was to vehicle), with empirical addition of chromoglycate, ranitidine, symbicort, monteleukast and prolongation of the protection period pre-infusion. We have no evidence of efficacy of any of these, but he has incrementally advanced to routine ward admissions for infusions, dose increase to 20 mg (0.3 mg/kg) Agalsidase-beta over 12 h, and administration of more convenient 36-h admissions at his local hospital each fortnight. He typically requires 1–3 doses of subcutaneous adrenaline (0.3 mg) for the reactions which characteristically occur 6–12 h after completion of most infusions.

Currently, he has been back on ERT for 2 years, has resumed sweating, is again coping with full-time work, and pain is controlled. Echocardiogram and GFR remain normal, and proteinuria is stable at 200 mg/day on ancillary therapy. Antibody testing remains negative for both IgG and IgE. Despite his ERT dose still being well below the approved dose of 1 mg/kg/fortnight, he has improved since infusions resumed. We plan to very slowly increase dose to standard levels, but will need to continue our empiric compromise between possible benefit of this, against extended infusion time, patient inconvenience and reaction risk.

Discussion

The male phenotype of Anderson–Fabry disease, although often clinically severe, is probably modified by ERT. In many countries, including the UK, North America, Europe and Australia, most affected patients can access ERT as part of their comprehensive medical care. Recombinant human GLA is commercially available in two forms, Agalsidase alfa (Replagal, Shire Human Genetic Therapies), and Agalsidase-beta (Fabrazyme, Genzyme)]), both administered intravenously. Each can induce infusion reactions, but excellent safety profiles in placebo controlled trials and open-label use are well documented (Eng et al. 2001a, b; Ramaswami et al. 2005; Barbey and Livio 2006). Reactions are generally easily manageable, reduce with time, and rarely interfere with ongoing ERT, although prophylaxis may be required (Wilcox et al. 2004; Barbey and Livio 2006). Agalsidase alfa infusion reactions are less common (approximately 13%) than those induced by Agalsidase-beta, possibly related to the lower protein dose administered in standard ERT (0.2 mg/kg vs. 1.0 mg/kg). Reactions are less common in patients with missense mutations, in whom low levels of endogenous enzyme can usually be detected.

Reactions very rarely necessitate ERT withdrawal. The typical pattern of Agalsidase-beta infusions were prospectively documented in Phase 3 and 4 trials undertaken in 58 patients over 30–36 months (Wilcox et al. 2004). IgG seroconversion occurred in 90% of patients at a median time of 6 weeks, and coincided in 70% of these with the onset of clinical infusion reactions. Reactions were typically mild, managed by infusion rate reduction with or without medication, and comprised various combinations of rigours, feeling cold or warm, fever, nausea, headache, and nasal congestion. Over time, infusion reactions decreased, as did antibody titre in most patients. Specific serum IgE or positive skin tests were found in three patients, all of whom were successfully rechallenged. Evidence of treatment efficacy was present independent of seroconversion, although subsequently Linthorst et al. (2004) reported failure of standard dose ERT to reduce urinary globotriaosylceramide in a subset of reacting patients with neutralising antibodies. In this study, IgG antibodies cross-reacted in vitro similarly with both recombinant enzymes, indicating that product switch is unlikely to be clinically useful.

Registry data from the Fabry Outcome Study (FOS) documented an incidence of reactions to Agalsidase alfa of 13% of all patients: 17% in males and 6% in females – representing 1% of total infusions over 900 patient years of infusions in 400 patients (Barbey and Livio 2006). Similar to Algalsidase beta, infusion reactions occurred in most affected patients soon after treatment initiation were easily managed by premedication (paracetamol, antihistamines and corticosteroids) and slowing of the infusion rate, and disappeared after the next few infusions. Reactions were severe in only 6% of patients experiencing reactions, and in only one patient was withdrawal from ERT required. No IgE antibodies were detected. IgG antibodies typically developed at approximately 3 months, but after 12–18 months of therapy 83% of patients treated with Agalsidase alfa were antibody free, and about 30% of antibody positive patients had developed immunological tolerance.

While ERT infusion reactions typically occur in male Fabry patients with a null mutation and specific IgG antibodies, their pathogenesis remains undefined. While the absolute protein infusion dose may explain the different rate of antibody formation between Agalsidase alpha and beta, their different glycosylation pattern may also be relevant (Schellekens 2002).

Our patient has a missense mutation, had only low titre IgG reverting to negative after desensitisation, and his delayed, atypical infusion reactions seem to be unrelated to antibody. It is interesting that none of five other family members have reacted to either form of long-term ERT. It is possible that his congenitally narrow airway and probable Fabry-related OSA exacerbated the ERT reactions, but the severity of reactions is unique in our experience. Stopping then resuming ERT is recognised as a risk factor for infusion reactions, and was possibly relevant to his reaction to his 13th infusion, occurring after a break of 3 months.

The unfortunate combination of rare, life-threatening disease and intractable ERT reactions promised him a progressive debilitating course, poor quality of life, ultimate organ failure and premature death. Demonstrating extraordinary determination to pursue treatment, he is again accessing ERT at a dose that has objective efficacy, albeit below "standard" dose. His currently tolerated fortnightly 0.3 mg/kg dose of Agalsidase-β will hopefully produce benefit worthy of his considerable efforts. To some extent, dosing of ERT is empiric – dose-finding and dose-comparator studies are very limited. Both forms of Fabry ERT have been shown – at their different "standard" doses – to have clinical efficacy. While their manufacturing processes, glycosylation patterns and individual organ uptakes are different, they have identical amino acid sequences, suggesting that – at least in some patients – there may be latitude for dose reduction of Agalsidase-beta.

We report our first patient in whom severe reactions to recombinant GLA necessitated treatment withdrawal. Subsequent clinical deterioration stimulated the trial of multiple therapeutic strategies, including successful rechallenge after tracheostomy.

Acknowledgements The authors gratefully acknowledge the persistence and courage of the patient and his family, the assistance of nursing and medical staff of the Nephrology, Otolaryngology, Intensive Care and Medical Day Units of the Royal Melbourne and the patient's regional hospitals, and the provision of Gb3 levels and antibody testing from the laboratories of Shire and Genzyme.

Synopsis

Tracheostomy allowed reinstitution of enzyme replacement therapy in a male Fabry patient, after withdrawal due to severe atypical infusion reactions despite extensive prophylaxis.

COI Disclosure

Dr Nicholls has received research and travel support from Genzyme and Shire HGT.

References

Barbey F, Livio F (2006) Safety of enzyme replacement therapy. In: Mehta A, Beck M, Sunder-Plassmann G (eds) Fabry disease: perspectives from 5 years of FOS. Oxford PharmaGenesis, Oxford, Chapter 41. http://www.ncbi.nlm.nih.gov/books/NBK11617/

Eng CM, Guffon N, Wilcox WR et al (2001a) Safety and efficacy of recombinant human α-galactosidase A – replacement therapy in Fabry's disease. N Engl J Med 345:9–16

Eng CM, Banikazemi M, Gordon RE et al (2001b) A phase 1/2 clinical trial of enzyme replacement in Fabry disease: pharmacokinetic, substrate clearance, and safety studies. Am J Hum Genet 68:711–722

Linthorst GE, Hollack CEM, Donker-Koopman WE, Strijland A, Aerts JMFG (2004) Enzyme therapy for Fabry disease: neutralizing antibodies toward agalsidase alfa and beta. Kidney Int 66:1589–1595

Ramaswami U, Wendt S, Parini R et al (2005) Safety of enzyme replacement therapy with agalsidase alfa in children with Fabry disease. J Inherit Metab Dis 28(Suppl 1):330

Schellekens H (2002) Bioequivalence and the immunogenicity of biopharmaceuticals. Nat Rev Drug Discov 1:457–462

Wilcox WR, Banikazemi M, Guffon N et al (2004) International Fabry Disease Study Group. Long-term safety and efficacy of enzyme replacement therapy for Fabry disease. Am J Hum Genet 75:65–74

JIMD Reports
DOI 10.1007/8904_2011_107

Infantile Progressive Hepatoencephalomyopathy with Combined OXPHOS Deficiency due to Mutations in the Mitochondrial Translation Elongation Factor Gene *GFM1*

S. Balasubramaniam · Y.S. Choy · A. Talib ·
M.D. Norsiah · L.P. van den Heuvel · R.J. Rodenburg

Received: 28 May 2011 /Revised: 17 October 2011 /Accepted: 20 October 2011 /Published online: 21 December 2011
© SSIEM and Springer-Verlag Berlin Heidelberg 2011

Abstract Mitochondrial disorders are a heterogeneous group of often multisystemic and early fatal diseases caused by defects in the oxidative phosphorylation (OXPHOS) system. Given the complexity and intricacy of the OXPHOS system, it is not surprising that the underlying molecular defect remains unidentified in many patients with a mitochondrial disorder. Here, we report the clinical features and diagnostic workup leading to the elucidation of the genetic basis for a combined complex I and IV OXPHOS deficiency secondary to a mitochondrial translational defect in an infant who presented with rapidly progressive liver failure, encephalomyopathy, and severe refractory lactic acidemia. Sequencing of the GFM1 gene revealed two inherited novel, heterozygous mutations: a.539delG (p.Gly180AlafsX11) in exon 4 which resulted in a frameshift mutation, and a second c.688G > A (p.Gly230Ser) mutation in exon 5. This missense mutation is likely to be pathogenic since it affects an amino acid residue that is highly conserved across species and is absent from the dbSNP and 1,000 genomes databases. Review of literature and comparison were made with previously reported cases of this recently identified mitochondrial disorder encoded by a nuclear gene. Although limited in number, nuclear gene defects causing mitochondrial translation abnormalities represent a new, rapidly expanding field of mitochondrial medicine and should potentially be considered in the diagnostic investigation of infants with progressive hepatoencephalomyopathy and combined OXPHOS disorders.

Communicated by: Shamima Rahman.

Competing interests: None declared.

S. Balasubramaniam (✉)
Genetics Department, Kuala Lumpur Hospital, Kuala Lumpur, Malaysia
e-mail: saras329@hotmail.com

S. Balasubramaniam
Metabolic Unit, Women's and Children's Hospital, Adelaide, SA 5006, Australia

Y.S. Choy
Prince Court Medical Centre, Kuala Lumpur, Malaysia

A. Talib
Department of Pathology, Kuala Lumpur Hospital, Kuala Lumpur, Malaysia

M.D. Norsiah
Molecular Diagnostics and Protein Unit, Institute for Medical Research, Kuala Lumpur, Malaysia

L.P. van den Heuvel · R.J. Rodenburg
Laboratory for Genetic, Endocrine and Metabolic Disorders, Department of Pediatrics, Nijmegen Center for Mitochondrial Disorders, Radboud University Nijmegen Medical Centre, Nijmegen, The Netherlands

L.P. van den Heuvel
Department of Pediatrics, Catholic University Leuven, Leuven, Belgium

Introduction

Mitochondria are ubiquitous intracellular organelles present in virtually all eukaryotic cells. They are relics of an ancestral alpha-proteobacterial endosymbiont (Gray et al. 1999) that took permanent residence in our cells. Genetic disorders of mitochondrial respiratory chain are the most common group of inborn errors of metabolism, collectively affecting approximately 1 in 5,000 births (Skladal et al. 2003). Mitochondrial dysfunction encompasses an extraordinary assemblage of clinical phenotypes, commonly

manifesting in tissues with high energy requirements, such as brain, retina, heart, muscle, liver, and endocrine systems (Cwerman-Thibault et al. 2011). It has also been implicated in a variety of diseases, including common multifactorial disorders such as diabetes (Gerbitz et al. 1996), Parkinson's disease (Mizuno et al. 1995; Mandemakers et al. 2007), and cancer (Brandon et al. 2006).

The complex and intricate nature of the oxidative phosphorylation (OXPHOS) system, which consists of about 90 proteins encoded by both the nuclear and the mitochondrial genome, explains the clinical heterogeneity associated with genetic defects in OXPHOS (Munnich et al. 1992). In this report, we describe an infant presenting with early fatal hepatoencephalopathy resulting from combined deficiencies of complex I and IV due to mutations in a nuclear gene encoding mitochondrial translational factor EFG1.

Case Report

The patient was the second child of a healthy, non-consanguineous Chinese couple, born at term via Cesarean section for intrauterine growth retardation and abnormal lie. The Apgar scores were good. Her weight of 2.03 kg and her head circumference of 30 cm were both below the third percentile, while her length was 49 cm, at the 25th percentile. On day 2 of life, she was noted to be lethargic, tachypnoeic, and hypoglycemic, with a venous blood sugar of 1.7 mmol/L. She had mild hyperammonemia, 167 μmol/L (normal <50), refractory raised anion gap metabolic acidosis (~22–30) with a pH of 6.8, greatly elevated serum lactate of 17–25 mmol/L (normal <2.4), and CSF lactate of 12 mmol/L (normal <2.1). She was ventilated and had single volume exchange transfusion followed by peritoneal dialysis performed in an effort to control the acidosis. Her urine organic acids showed excessive excretion of lactate, 3-OH butyric, 2-OH butyric acids, 4-OH phenyllactate, and ketonuria. Her acylcarnitines were normal and her serum amino acids revealed elevated alanine at 1,228 μmol/L (normal 122–546), glutamine 1,114 μmol/L (normal 59–561), methionine 128 μmol/L (normal 10–79), phenylalanine 165 μmol/L (normal 31–157), and tyrosine 569 μmol/L (normal 5–167). Liver function test showed hypoproteinemia (protein 42 and albumin 24 g/L), raised alkaline phosphatase 516 IU/L (age-related reference range 0–341), gamma-glutamyl transferase 523 IU/L (normal 11–50), mildly elevated alanine aminotransferase 66 U/L (normal 0–54), aspartate aminotransferase 86 U/L (normal 0–82), and total bilirubin of 108 μmol/L (normal 0–17). She responded to peritoneal dialysis with normalization of acidosis and a reduction of serum lactate to 6 mmol/L. Echocardiogram was normal and ultrasound brain showed dilated ventricles. She had notable

generalized hypotonia with myopathic facies and subtle dysmorphism including flat nasal bridge, low-set ears, high, broad forehead, and smooth philtrum. She was suspected to have a possible mitochondrial DNA (mtDNA) depletion syndrome and commenced on oral Coenzyme Q 10, oral thiamine, riboflavin, biotin, and vitamin E. She was discharged home at 2 weeks of age on breast feeding with normal blood gas and lactate.

Oral sodium bicarbonate was supplemented at 4 weeks of life with the recurrence of systemic acidosis and lacticacidemia. She developed persistent vomiting and steatorrhoea. Clinically an enlarged liver was palpable 5 cm below the right costal margin. Progressive deterioration of liver function occurred with conjugated hyper-bilirubinemia (direct bilirubin 70 μmol/L), total 134 (normal 0–17), worsening hypoalbuminemia, and transaminemia.

In the following months, she remained hypotonic with failure to thrive and globally delayed developmental milestones. She was readmitted at 10 weeks of age with escalating lethargy and inactivity precipitated by recurrent vomiting and loose stools. Clinically, she demonstrated features of circulatory compromise with cold, cyanosed peripheries. Biochemical parameters included markedly raised lactate at 15 mmol/L and compensated metabolic acidosis (pH 7.35, bicarbonate 12.7, base excess −13). Brain MRI showed global cystic changes in the subcortical white matter, T2 hyperintensities in the putamen, globus pallidi, and ventricular dilatation with septatation (Fig. 1a–c). She was discharged home after several days with anti-reflux infant formula, oral ranitidine, and domperidone.

Shortly thereafter, she was readmitted with a similar episode of recurrent vomiting and systemic acidosis, and was subsequently referred to a tertiary center for further management. Features noted at initial assessment here showed an emaciated, pale, and mildly jaundiced infant with a weight of only 2.9 kg at 4 months of age. She had reduced spontaneous movements, hypotonia with severe head lag, myopathic facies, and reduced deep tendon reflexes. Intermittent roving nystagmus and mild ptosis were also noted. A soft ejection systolic murmur was auscultated over her upper left sternal edge. Echocardiography showed a small, 0.21 cm, atrial septal defect, with good contractility and ejection fraction of 65.8%. Her liver was palpable 6 cm below the right costal margin. Her hemoglobin was 9.5 g/dL with normochromic normocytic cells on film and adequate reticulocyte response. The other cell lines were normal. Hepatic dysfunction was evident with mild coagulopathy; activated partial thromboplastin time 50 s (control 30.9–45.9 s), prothrombin time 19 s (control 11.9–14 s), serum albumin 27 g/L (normal 30–54), and mildly elevated transaminases; alanine aminotransfer-

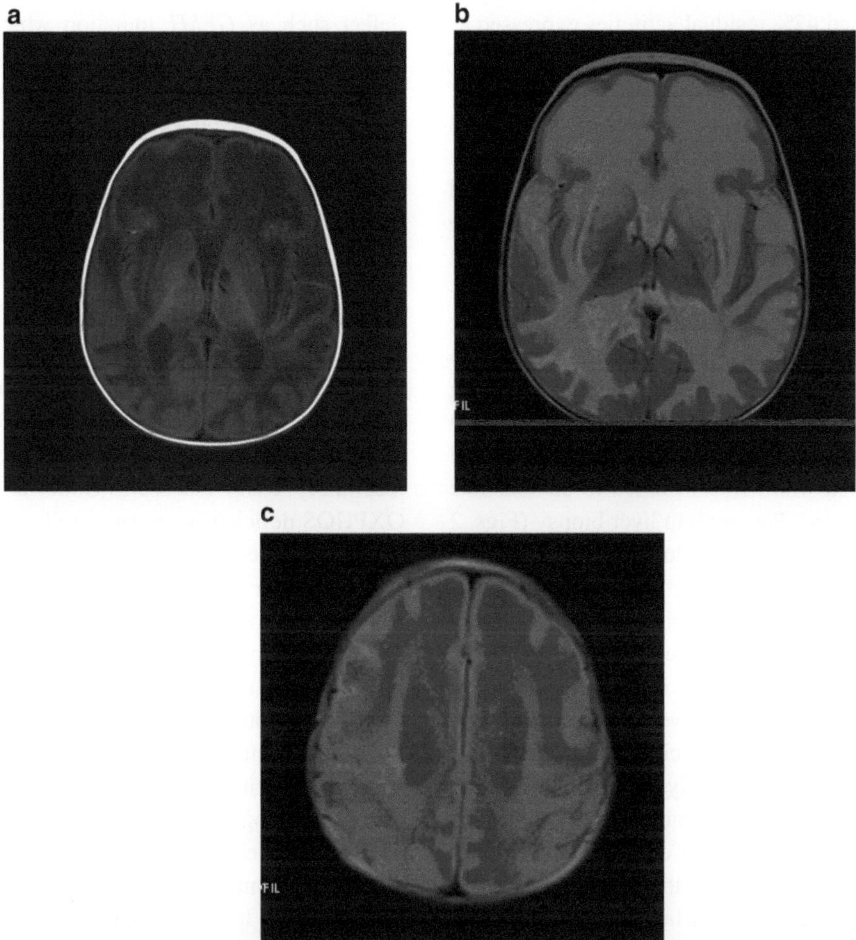

Fig. 1 (**a**, **b**) MRI brain T1 and T2 weighted axial images showing global, cystic changes and hyperintensities in putamen and glubus pallidi. (**c**) MRI brain showing subcortical cystic changes and dilated lateral ventricles with ventricular septae

ase 170 U/L (normal 0–33), aspartate aminotransferase 116 (normal 0–82). Total bilirubin was 115 μmol/L (normal 0–17) with conjugated bilirubin 90 μmol/L, raised alkaline phosphatase 954 U/L (age-related reference range 0–356), gamma-glutamyl transferase 289 U/L (normal 11–50), and ammonia 65 μmol/L (normal <50). Serum lactate was 17.7 mmol/L, pyruvate 445.6 μmol/L (normal 30–80), blood pH 7.151, pCO_2 12.1 mmHg, pO_2 168.8 mmHg, bicarbonate 4, and base excess −24.5. Serum creatine kinase, very long chain fatty acids, carnitine, and transferrin isoforms were normal. Plasma alanine and proline were raised at 1,116 μmol/L (normal 132–455) and 508 μmol/L (normal 77–329), respectively. Urine organic acids showed increased excretion of lactate, ketones, citramalic, and fumaric acid.

Abdominal ultrasound showed a left multicystic kidney with a dilated distal left ureter and enlarged right kidney with bipolar length of 5.5 cm. The liver was enlarged with a smooth outline and absence of any focal lesions. Repeat brain ultrasound revealed multiple cystic changes predominantly observed in bifrontal lobes, and periventric-

ular regions with ex-vacuo dilatation of the lateral ventricles. Her thyroid function studies were abnormal; free T4 12.1 pmol/L (14.1–19.2), TSH 7.55 mU/L (0.98–5.3). Ophthalmological assessment reported the presence of alternating exotropia with poor fixation, but normal fundus, possibly indicating cortical visual impairment. Treatment included attempts to correct electrolyte imbalances, metabolic derangements with oral sodium bicarbonate, and addressing feeding difficulties with nasogastric feeding of anti-reflux infant formula with supplemented medium chain triglycerides (MCT). Nevertheless her weight failed to pick up and at 8+ months of age, she only weighed 4.2 kg, well below the third percentile. She developed increasingly severe liver impairment, systemic metabolic and lactic acidosis. Her parents opted for a conservative management and she succumbed at home from respiratory failure at 8 months of age.

Meanwhile, further investigations were carried out to identify the precise cause of her illness. Biochemical examination cultured fibroblasts had demonstrated reduced respiratory chain enzyme activities (OXPHOS) of complex

I and IV with 68% and 47% residual activities expressed as a percent of the lowest control value respectively. The other OXPHOS enzymes complex II, complex III, and complex V showed a normal activity. Assays to quantify OXPHOS enzyme activities were based on spectrophotometry (Janssen et al. 2007; Mourmans et al. 1997; Cooperstein and Lazarow 1951; Jonckheere et al. 2008; Srere 1969). Muscle and liver OXPHOS assays would perhaps have yielded more representative results of the tissue-specific involvement observed in mitochondrial respiratory chain disorders; however, her parents had declined these tests due to concerns over her fragile state. Molecular genetic testing of a total of 12 point mutations in mtDNA isolated from blood for five genes associated with Leigh disease and seven genes associated with MELAS failed to demonstrate any pathogenic mutations. Postmortem liver biopsy (Figs. 2 and 3) and histopathology displayed lobular disarray, portal fibrosis, micro and macrovesicular steatosis, and intrahepatic and intracanicular cholestasis. Due to financial constraints, we were unable to analyze OXPHOS assays on postmortem liver tissue or perform direct sequencing of mtDNA. mtDNA depletion studies were not performed as the finding of reduced enzyme deficiencies in fibroblasts make a depletion syndrome less likely. Her parents had not consented for postmortem muscle biopsy. The associated biochemical evidence of a combined enzyme deficiency in cultured fibroblasts, in addition to the early, severe, and rapid progression of a predominant hepatocerebral disease in our patient, suggested a possible candidate nuclear gene

defect such as *GFM1* mutation which had only recently been described in a handful of patients with a similar phenotype. Subsequent sequencing of *GFM1* gene was performed (Smits et al. 2010a) and revealed the heterozygous changes c.539delG (p.Gly180AlafsX11) in exon 4 which was inherited from the father, and c.688G > A (p. Gly230Ser) in exon 5, inherited from the mother.

Discussion

OXPHOS disorders may occur as isolated enzyme deficiencies and have been reported in approximately 67% cases, while combined enzyme deficiencies account for 33% of all respiratory chain disorders (Smits et al. 2010b). Isolated OXPHOS deficiencies are generally caused by mutations in structural genes which encode subunits of the OXPHOS system, or in genes encoding proteins involved in the assembly of a specific OXPHOS enzyme complex (Zeviani and Di Donato 2004). Combined OXPHOS defects tend to involve genes required for mtDNA maintenance, mitochondrial transcription, or translation including posttranscriptional or posttranslational processes, import of nDNA-encoded proteins into the mitochondrion or mitochondrial membrane biogenesis (Smits et al. 2010b). Approximately 40% of all combined respiratory chain deficiencies occur as a result of mtDNA deletions and point mutations in mitochondrial transfer RNA genes, which more frequently affect adult patients (Kemp et al. 2011). Another 40% of

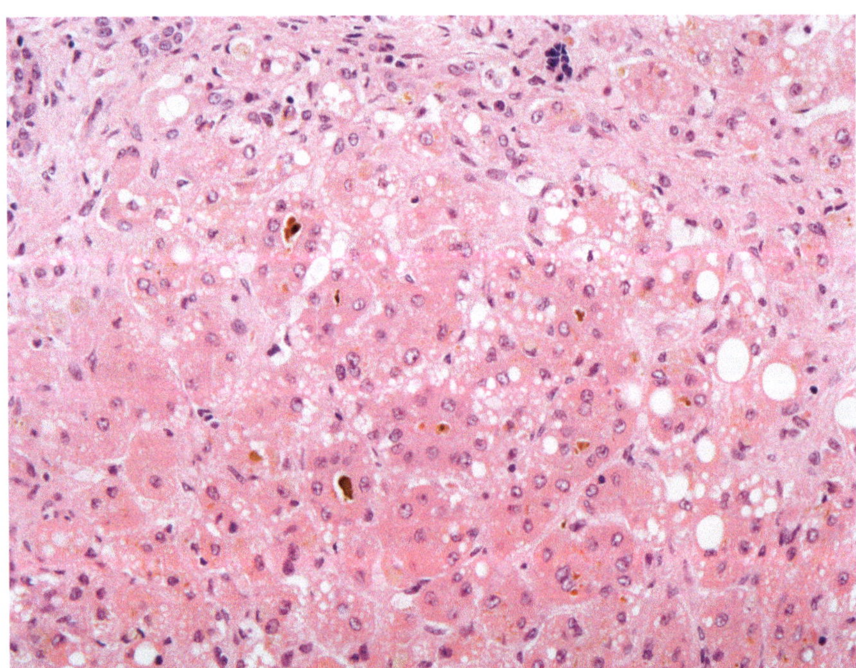

Fig. 2 Liver H&E X40: Liver tissue displaying moderate microvesicular steatosis and patchy macrovesicular steatosis with intrahepatic and intracanicular cholestasis. The portal tracts are expanded with accompanying portal fibrosis and portal to portal bridging fibrosis

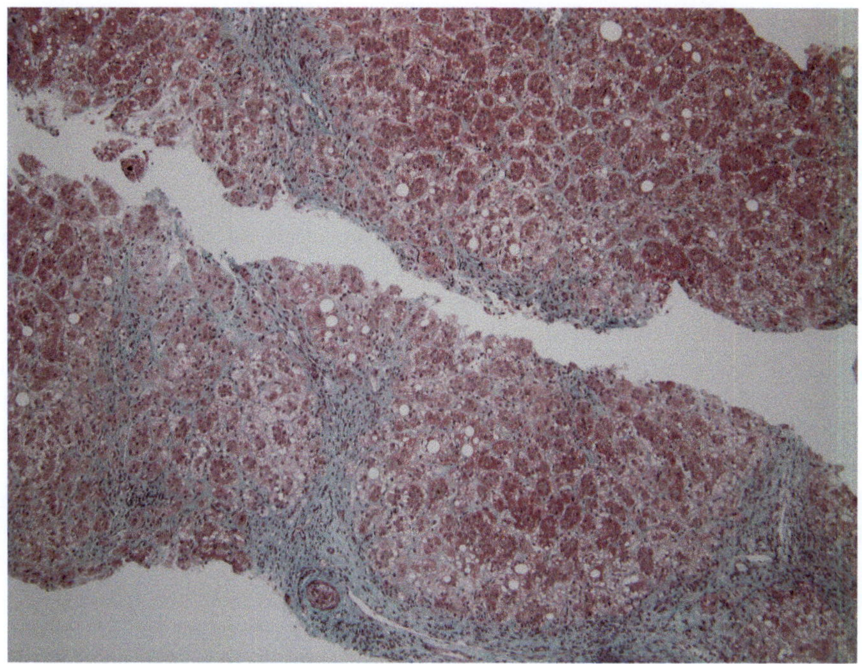

Fig. 3 Liver MT X10: Lobular disarray of liver tissue is evident with the nodular pattern emphasized by Masson's Trichrome stain

combined OXPHOS deficiencies are related to mtDNA depletion, which predominantly affect young children and are caused by autosomal recessive mutations in nuclear genes (*DGUOK, MPV17, POLG, TYMP, TK2, SUCLA2, SUCLG1, RRM2B, PEO1*) influencing mtDNA replication and maintenance (Spinazzola et al. 2009). In the remaining 20% of combined respiratory chain deficiencies, after excluding mtDNA deletions, depletion, and point mutations, no clear diagnostic pathway is currently available to determine the cause of disease. The reduced enzyme activities in our patient's cultured fibroblasts made the possibility of mtDNA depletion less likely as it has been reported that exponentially growing cells may not manifest any OXPHOS enzyme deficiencies, but instead have to be kept in a quiescent state in order to clearly demonstrate their phenotype (Pontarin et al. 2011; Gonzalez-Vioque et al. 2011). For several types of depletion syndromes, it has been shown that nucleotide metabolism plays an important role in this phenomenon (Gonzalez-Vioque et al. 2011). We used growing fibroblasts with a low passage number, and under these conditions fibroblasts from depletion syndrome patients usually do not show OXPHOS enzyme deficiencies.

A thorough clinical characterization of patients is imperative in identifying homogeneous patient groups and analyzing the complex molecular mechanisms behind combined respiratory chain deficiencies. When facilitated by linkage studies in consanguineous families and functional cell culture investigations conducted at various regulatory levels of mitochondrial function, transcription,

translation, ribosome function, protein stability, subcomplex formation, it may provide a potentially promising approach in selecting novel candidates in combined respiratory chain deficiencies (Kemp et al. 2011).

Eukaryotes contain two translational systems, one in the cytosol and the other in the mitochondria. The mitochondrial translation machinery comprises mtDNA-encoded rRNAs and tRNAs, in addition to various proteins encoded by the nuclear genome including two initiation factors, IF2 (*MTIF*) (Ma and Spremulli 1995), IF3 (*MTIF3*) (Koc and Spremulli 2002); three elongation factors, EFTu (*TUFM*) (Ling et al. 1997), EFTs (*TSFM*) (Xin et al. 1995), EFG1 (*GFM1*) (Gao et al. 2001); four release factors, RF1 (*MTRF1*) (Zhang and Spremulli 1998), RF1a (*HMRF1L*) (Zhang and Spremulli 1998), *C12ORF65* (Antonicka et al. 2010), *ICT1* (Richter et al. 2010); and two recycling factors, RRF (*MRRF*) (Soleimanpour-Lichaei et al. 2007) and EFG2, which has been renamed RRF2 (*GFM2*) (Zhang and Spremulli 1998; Rorbach et al. 2008; Tsuboi et al. 2009); mitochondrial ribosomal proteins (MRPs); mitochondrial aminoacyl-tRNA synthetases, and methionyl-tRNA transformylase (Smits et al. 2010b). Although most components of the mitochondrial translation system are nuclear encoded, the majority of patients harboring mutations in nuclear genes are very limited. Most mutations associated with mitochondrial protein synthesis defects to date have been reported in mtDNA.

Recently, mutations in *GFM1*, previously known as *EFG1*, a nuclear factor of the mitochondrial translation machinery, have been described (Smits et al. 2010a;

Coenen et al. 2004; Antonicka et al. 2006; Valente et al. 2007). Mitochondrial *GFM1* is a five-domain GTPase that catalyzes the translocation step of mitochondrial protein synthesis, during which peptidyl-tRNA moves from the ribosomal acceptor (amino acyl or A) site to the ribosomal peptidyl (P) site following removal of the deacylated tRNA from the ribosomal (P) site to the exit (E) site. This results in the concomitant advancement of the mRNA by one codon and exposure of the next codon in the A site preparing for a new elongation cycle (Wintermeyer et al. 2004). The compilation of biochemical evidence of a combined enzyme deficiency in cultured fibroblasts of complex I and IV, which is a clear and well-known functional consequence of *GFM1* mutations (Smits et al. 2010a; Coenen et al. 2004; Antonicka et al. 2006; Valente et al. 2007), in addition to the severe and early onset presentation with rapidly progressive hepatocerebral disease in our patient suggested a possible candidate nuclear gene defect such as *GFM1* mutation which had only recently been described in a handful of patients with a predominant hepatic and/or neurological phenotype.

Sequence analysis of the complete coding region of the *GFM1* gene in our patient revealed the heterozygous changes c.539delG (p.Gly180AlafsX11) in exon 4 which was inherited from the father had resulted in a frameshift mutation, while a second c.688G > A (p.Gly230Ser) mutation in exon 5 was inherited from the mother. The pathogenicity of the second novel missense mutation was suggested by its highly conserved nature in all nine organisms of which a *GFM1* sequence could be found, and its absence in the dbSNP database. In addition, both these mutations were not present in the 1,000 genomes database which includes 60 Chinese genomes and 194 South-East Asian genomes.

Previously, four other *GFM1* defects have been described. A review of their clinical features and investigations has been tabulated in Table 1. The first report of two siblings of consanguineous Lebanese parents harbored homozygous mutations, c.521A > G in the GTP-binding domain I (Coenen et al. 2004). The index patient had presented with intrauterine growth retardation, mild microcephaly, hypertonicity with reduced spontaneous movements, and she soon developed profound metabolic acidosis and lactic academia with raised lactate: pyruvate ratio of 38 (normal 12–18) by day 10 of life. Rapidly progressing liver failure occurred leading to death on day 27 of life.

Antonicka and colleagues described similar clinical phenotypes in two siblings with significant growth retardation, lactic acidosis, fatal hepatopathy, small or undeveloped corpus callosum, and severe respiratory deficiency in skeletal muscle and liver, without any evidence of clinical cardiac dysfunction (Antonicka et al. 2006). The molecular basis for the novel mutations in *GFM1* demonstrated that the translation defect resulting from these mutations was associated with unique, tissue-specific patterns of OXPHOS deficiency. The severity of the OXPHOS defect correlates with the residual levels of the mutant EFG1 protein in different tissues. In addition, there appears to be an adaptive response that involves transcriptional upregulation of EFTu, another translation elongation factor, with significant quantitative differences in the ratios of the translation elongation factors. This may reasonably reflect the different tissue-specific demands for the mitochondrially encoded polypeptides and explain how the heart uniquely possesses mechanisms to circumvent the disruption of mitochondrial translation caused by mutations in the nuclear-encoded components of the translation machinery.

A subsequent report by Valente and colleagues on an infant affected by neonatal lactic acidosis, rapidly progressive encephalopathy, severely decreased mitochondrial protein synthesis, and combined OXPHOS deficiency was a compound heterozygote for two novel *GFM1* mutations (Valente et al. 2007). No data are available concerning the expression of mitochondrial *GFM1* in human brain, but the predominant neurological involvement with neuroradiological hallmarks of early onset Leigh syndrome including bilateral necrotizing lesions in the basal ganglia and brain stem overwhelmed, that of other organs notably, skeletal muscle, liver, and heart in this patient. It is postulated that the clinical variability may be attributed to various mechanisms, including the different sites of mutations in the EFG1 protein which may subsequently have tissue-specific effects on mitochondrial translation, adaptive processes (e.g., compensatory changes in other translational factor) which may act differently in different patients, and finally partial compensation by the second isoform of mitochondrial *EFG*, namely *EFG2* which is highly expressed in energy-consuming tissues, such as skeletal muscle, heart, and fetal liver. The functional role of *EFG2*, however, is uncertain (Smits et al. 2010a).

Strikingly, in the recent report by Smits et al., a mutation in *GFM1* was observed in a patient affected by severe, rapidly progressive mitochondrial encephalopathy; the decrease in enzyme activities of complex I, III, and IV detected in fibroblasts was not found in muscle tissue (Smits et al. 2010a). Reduced respiratory chain activities in fibroblasts with normal values in muscle tissue are an uncommon finding in mitochondrial disorders, whereas the opposite is often observed. The capacity of the mitochondrial energy-generating system, however, was clearly reduced in muscle tissue, indicated by impairments in pyruvate oxidation and ATP production rates. This evinces the importance of a thorough diagnostic biochemical analysis of muscle tissue and fibroblasts in patients as OXPHOS defects may be selectively expressed.

The clinical symptomatology in our patient is reminiscent of the severe phenotype of the first two reports

Table 1 Clinical features, biochemical analysis of OXPHOS assay on liver tissue, skeletal muscle and fibroblasts, mutation testing of *GFM1* gene and neuroradioimaging of the six patients diagnosed to date in comparison to our patient

	Clinical features	OXPHOS assay on fibroblasts	OXPHOS assay on liver tissue	OXPHOS assay on muscle tissue	*GFM1* mutation (cDNA level)	Brain imaging
Patient 1[a]	Female index patient, born to consanguineous parents with intrauterine growth retardation, mild microcephaly, and hypertonicity. Rapidly progressive liver failure with initial liver dysfunction from day 7, profound metabolic and lactic acidosis from day 10, and early death on day 27 of life.Blood lactate – 17.1 mmol/L, Lactate:pyruvate-38	Complex I – 40% Complex III – 69% Complex IV – 18%	Not done	Complex I – 52% Complex IV – 54%	Homozygousc.521A > G mutation	MRI-hypoplasia of the corpus callosum with several symmetrical cystic lesions in the white matter in the area of the basal ganglia
Patient 2[a]	Male sibling with extremely delayed growth and development, as well as increased muscle tone in his upper extremities. Signs of liver failure were present at week 7, leading to death at 5 months of ageBlood lactate – 9.3 mmol/L, Lactate:pyruvate-84	Complex I – 13% Complex IV – 31%	Not done	Not done	Homozygousc.521A > G mutation	Ultrasound brain demonstrated generalized atrophy and a small corpus callosum
Patient 3[b]	Female neonate with growth retardation, lactic and metabolic acidosis, rapidly progressive fatal hepatopathy on day 9 of life from pulmonary hemorrhage. Patent ductos arteriosus presentBlood lactate – 34 mmol/L	Complex I – 20% Complex III – 40–60% Complex IV – 20% Complex V – 40–60%	Complex I <10% Complex IV < 10% Complex V – 50%	Complex I – 50% Complex IV – 20% Complex V – 20%	c.1068T > C; c.1872-2 del AG	Brain MRI was normal on day 5 but MRS showed marked elevation of lactate
Patient 4[b]	Termination of pregnancy for IUGR and significant oligohydramnios. The baby died 45 min after birth. Amniocentesis 15 weeks – 45X (4)/46XX (16) karyotype in cultured cells, with inconclusive analysis of lactate:pyruvate ratio, COX activity, immunocytochemistry, and western blots	Complex I – 30% Complex III – 60% Complex IV – 10% Complex V – 50%	Not done	Not done	Not done	Not done
Patient 5[c]	Dysmorphic features, including flat nasal bridge, low-set ears, small hands and feet, epicanthus, and high, arched palate noted at day 7 of life. At age 3 weeks, she started having feeding difficulties, weight loss, and subsequently had rapidly progressive encephalopathy. She passed away at 16 months from respiratory insufficiency	Complex I – 10% Complex II – 25% Complex IV – 53%	Not done	Complex I – 4% Complex II – 17% Complex III – 58% Complex IV – 17% Complex V – 35%	c.139C > T; c.1478T > G	MRI brain showed two large, bilateral, and symmetrical areas of increased T2 involving the putamen and the globus pallidus

Table 1 (continued)

	Clinical features	OXPHOS assay on fibroblasts	OXPHOS assay on liver tissue	OXPHOS assay on muscle tissue	*GFM1* mutation (cDNA level)	Brain imaging
Patient 6[d]	Female patient, second of a dizygotic twin, small for gestational age presented at day 2 of life with feeding problems, encephalopathy, and hypotonia. Seizure onset at 8 weeks of life with microcephaly, delayed visual maturation. Progressive neurological deterioration occurred and succumbed at 2 years of age from pneumoniaBlood lactate – 4.9 mmol/L, Lactate:pyruvate-22	Complex I – 34 mU/U CS (reference range 100–310 mU/U CS)Complex III – 1,038 mU/U CS (reference range 1,320–2,610 mU/U CS) Complex IV – 168 mU/U CS (reference range 1,320–2,610 mU/U CS)	Not done	Complex III – 2,151 mU/U CS (reference range 2,200–6,610 mU/U CS)	Homozygous c.748C > T	Brain MRI revealed a small frontal cortex, a thin corpus callosum and delayed myelination
Patient 7[e]	Intrauterine growth retardation, severe metabolic and lactic acidosis day 2 of life, with progressive liver failure. Some dysmorphic features noted, atrial septal defect, abnormal kidneys	Complex I – 68% 112 mU/U CS (reference range 163–599 mU/U CS)Complex IV – 47% 136 mU/U CS (reference range 288–954 mU/U CS)			c.539delG; c.688G > A	Brain MRI showed globally extensive cystic changes in subcortical white matter. Increased T2 of putamen, globus pallidi, and ventricular dilatation with septations

[a] Siblings described by Coenen et al. 2004

[b] Siblings described by Antonicka et al. 2006

[c] Valente et al. 2007

[d] Smits et al. 2010a

[e] Our patient

(Coenen et al. 2004; Antonicka et al. 2006), characterized by intrauterine growth retardation, profound lactic acidosis, and progressive liver dysfunction, resulting in liver failure and death within the first weeks or months. In addition, our patient had a multicystic kidney and an atrial septal defect. Neuroradiological evidence of cystic changes in temporal lobes with white matter abnormalities and biochemical findings of high blood lactate levels of 25 mmol/L and lactate: pyruvate ratios of 38 suggest a more global and deleterious disturbance of mitochondrial functioning.

In summary, unique patterns of OXPHOS deficiency result when one of the nuclear-encoded components of the translation system fail. The clinical heterogeneity associated with mutations in the mitochondrial translation apparatus may reflect different tissue-specific demands for the mitochondrially encoded polypeptides with significant quantitative differences in the ratios of the translation elongation factors. Despite rapid advancements in our understanding of the mechanisms implicated in mitochondrial disease, it is far from complete. The intricacy and complexities of the OXPHOS system renders the identification of the genetic defect an arduous task. Rapid advances in technologies involving high-throughput sequencing which are much cheaper and faster than the conventional approach of polymerase chain reaction followed by capillary sequencing (Tucker et al. 2009) or exome sequencing (Ng et al. 2009) which is the targeted sequencing of all protein-coding regions when applied with appropriate bioinformatics tools potentially offer a promising approach in elucidating the genetic etiology of OXPHOS deficiencies. Clearly, much work remains to be done to fully comprehend the processes involved in mitochondrial translation and the biogenesis of the OXPHOS system and their roles in the pathogenesis of combined OXPHOS deficiencies.

Acknowledgments The authors would like to express their sincere gratitude to the Director General of Health, Ministry of Health, Malaysia for allowing the publication of this case report.

Contributions of Authors

1. Balasubramaniam S – Clinical management of the patient, draft of manuscript, and completed version.
2. YS Choy – Clinical management of patient, review of the draft, and contribution to the completed article.
3. Talib A – Histopathological analysis of liver biopsy, review of the draft, and contribution to the completed article.
4. Norsiah MD – Molecular testing of parental DNA for GFM1 mutation, review of the draft, and contribution to the completed article.

5. van den Heuvel LP – Biochemical analyses of OXPHOS assay and molecular testing in the proband, review of the draft, and contribution to the completed article.
6. Rodenburg RJ – E Biochemical analyses of OXPHOS assay and molecular testing in the proband, review of the draft, and contribution to the completed article.

One Sentence Take Home Message

This report demonstrates a combined OXPHOS deficiency detected in patient fibroblasts, occurring as a result of a nuclear-encoded mitochondrial translational defect secondary to *GFM1* mutation, ultimately leading to mitochondrial hepatoencephalomyopathy and death at 8 months of age.

All authors declare that the answers to all questions on the JIMD competing interest form are "no" and therefore have nothing to declare.

References

Antonicka H, Sasarman F, Kennaway NG et al (2006) The molecular basis for tissue specificity of the oxidative phosphorylation deficiencies in patients with mutations in the mitochondrial translation factor EFG1. Hum Mol Genet 15:1835–1846

Antonicka H, Ostergaard E, Sasarman F, Weraarpachai W, Wibrand F, Anne Pedersen AM, Rodenburg RJ, van der Knaap MS, Smeitink JA, Chrzanowska-Lightowlers AM, Shoubridge EA (2010) Mutations in C12orf65 in patients with encephalomyopathy and a mitochondrial translation defect. Am J Hum Genet 87:115–122

Brandon M, Baldi P, Wallace DC (2006) Mitochondrial mutations in cancer. Oncogene 25:4647–4662

Coenen MJ, Antonicka H, Ugalde C et al (2004) Mutant mitochondrial elongation factor G1 and combined oxidative phosphorylation deficiency. N Engl J Med 351:2080–2086

Cooperstein SJ, Lazarow A (1951) A microspectrophotometric method for the determination of cytochrome oxidase. J Biol Chem 189(2):665–670

Cwerman-Thibault H, Sahel JA, Corral-Debrinski M (2011) Mitochondrial medicine: to a new era of gene therapy for mitochondrial DNA mutations. J Inherit Metab Dis 34(2):327–344

Gao J, Yu L, Zhang P et al (2001) Cloning and characterization of human and mouse mitochondrial elongation factor G, GFM and Gfm, and mapping of GFM to human chromosome 3q25.1-q26.2. Genomics 74:109–114

Gerbitz KD, Gempel K, Brdiczka D (1996) Mitochondria and diabetes: genetic, biochemical, and clinical implications of the cellular energy circuit. Diabetes 45(2):113–126

Gonzalez-Vioque E, Torres-Torronteras J, Andreu AL, Mart R (2011) Limited dCTP availability accounts for mitochondrial DNA depletion in mitochondrial neurogastrointestinal encephalomyopathy (MNGIE). PLoS Genet 7(3):e1002035

Gray MW, Burger G, Lang BF (1999) Mitochondrial evolution. Science 283(5407):1476–1481

Janssen AJ, Trijbels FJ, Sengers RC et al (2007) Spectrophotometric assay for complex I of the respiratory chain in tissue samples and cultured fibroblasts. Clin Chem 53:729–734

Jonckheere AI, Hogeveen M, Nijtmans LG, van den Brand MA, Janssen AJ, Diepstra JH, van den Brandt FC, van den Heuvel LP,

Hol FA, Hofste TG, Kapusta L, Dillmann U, Shamdeen MG, Smeitink JA, Rodenburg RJ (2008) A novel mitochondrial ATP8 gene mutation in a patient with apical hypertrophic cardiomyopathy and neuropathy. J Med Genet 45(3):129–133

Kemp J, Smith P, Pyle A et al (2011) Nuclear factors involved in mitochondrial translation cause a subgroup of combined respiratory chain deficiency. Brain 134:183–195

Koc EC, Spremulli LL (2002) Identification of mammalian mitochondrial translational initiation factor 3 and examination of its role in initiation complex formation with natural mRNAs. J Biol Chem 277:35541–35549

Ling M, Merante F, Chen HS et al (1997) The human mitochondrial elongation factor tu (EF-Tu) gene: cDNA sequence, genomic localization, genomic structure, and identification of a pseudogene. Gene 197:325–336

Ma L, Spremulli LL (1995) Cloning and sequence analysis of the human mitochondrial translational initiation factor 2 cDNA. J Biol Chem 270:1859–1865

Mandemakers W, Morais VA, De Strooper B (2007) A cell biological perspective on mitochondrial dysfunction in Parkinson disease and other neurodegenerative diseases. J Cell Sci 120 (10):1707–1716

Mizuno Y, Ikebe S, Hattori N et al (1995) Role of mitochondria in the etiology and pathogenesis of Parkinson's disease. Biochim Biophys Acta 1271:265–274

Mourmans J, Wendel U, Bentlage HA, Trijbels JM, Smeitink JA, de Coo IF, Gabreëls FJ, Sengers RC, Ruitenbeek WJ (1997) Clinical heterogeneity in respiratory chain complex III deficiency in childhood. J Neurol Sci 149(1):111–117

Munnich A, Rustin P, Rötig A et al (1992) Clinical aspects of mitochondrial disorders. J Inherit Metab Dis 15(4):448–455

Ng SB, Turner EH, Robertson PD et al (2009) Targeted capture and massively parallel sequencing of 12 human exomes. Nature 461 (7261):272–276

Pontarin G, Ferraro P, Rampazzo C et al (2011) Deoxyribonucleotide metabolism in cycling and resting human fibroblasts with a missense mutation in p53R2, a subunit of ribonucleotide reductase. J Biol Chem 286(13):11132–11140

Richter R, Rorbach J, Pajak A, Smith PM, Wessels HJ, Huynen M, Smeitink JA, Lightowlers RN, Chrzanowska-Lightowlers ZM (2010) A functional peptidyltRNA hydrolase, ICT1, has been recruited into the human mitochondrial ribosome. EMBO J 29:1116–1125

Rorbach J, Richter R, Wessels HJ, Wydro M, Pekalski M, Farhoud M, Kuhl I, Gaisne M, Bonnefoy N, Smeitink JA, Lightowlers R, Chrzanowska-Lightowler ZM (2008) The human mitochondrial ribosome recycling factor is essential for cell viability. Nucleic Acids Res 36(18):5787–5799

Skladal D, Halliday J, Thorburn DR (2003) Minimum birth prevalence of mitochondrial respiratory chain disorders in children. Brain 126:1905–1912

Smits P, Antonicka H, van Hasselt PM et al (2010a) Mutation in subdomain G' of mitochondrial elongation factor G1 is associated with combined OXPHOS deficiency in fibroblasts but not in muscle. Eur J Hum Genet. doi:10.1038/ejhg.2010.208

Smits P, Smeitink J, van den Heuvel L (2010) Mitochondrial translation and beyond: processes implicated in combined oxidative phosphorylation deficiencies. J Biomed Biotechnol (Article ID 737385, 24 pages, Epub 2010 Apr 13). doi:10.1155/2010/737385

Soleimanpour-Lichaei HR, Kuhl I, Gaisne M, Passos JF, Wydro M, Rorbach J, Temperley R, Bonnefoy N, Tate W, Lightowlers R, Chrzanowska-Lightowler ZMA (2007) mtRF1a is a human mitochondrial translation release factor decoding the major termination codons UAA and UAG. Mol Cell 27:745–757

Spinazzola A, Invernizzi F, Carrara F et al (2009) Clinical and molecular features of mitochondrial DNA depletion syndromes. J Inherit Metab Dis 32:143–158

Srere PA (1969) Citrate synthase, EC 4.1.3.7 citrate oxaloacetate lyase (CoA-acetylating). Methods Enzymol 13:3–11

Tsuboi M, Morita H, Nozaki Y, Nozaki Y, Akama K, Ueda T, Ito K, Nierhaus KH, Takeuchi N (2009) EF-G2mt is an exclusive recycling factor in mammalian mitochondrial protein synthesis. Mol Cell 35(4):502–510

Tucker T, Marra M, Friedman JM (2009) Massively parallel sequencing: the next big thing in genetic medicine. Am J Hum Genet 85(2):142–154

Valente L, Tiranti V, Marsano RM et al (2007) Infantile encephalopathy and defective mitochondrial DNA translation in patients with mutations of mitochondrial elongation factors EFG1 and EFTu. Am J Hum Genet 80:44–58

Wintermeyer W, Peske F, Beringer M et al (2004) Mechanisms of elongation on the ribosome: dynamics of a macromolecular machine. Biochem Soc Trans 32:733–737

Xin H, Woriax V, Burkhart W et al (1995) Cloning and expression of mitochondrial translational elongation factor Ts from bovine and human liver. J Biol Chem 270:17243–17249

Zeviani M, Di Donato S (2004) Mitochondrial disorders. Brain 127 (10):2153–2172

Zhang Y, Spremulli LL (1998) Identification and cloning of human mitochondrial translational release factor 1 and the ribosome recycling factor. Biochim Biophys Acta 1443:245–250

JIMD Reports
DOI 10.1007/8904_2011_114

CASE REPORT

Expanding the Spectrum of PMM2-CDG Phenotype

Sandrine Vuillaumier-Barrot · Bertrand Isidor ·
Thierry Dupré · Christiane Le Bizec · Albert David ·
Nathalie Seta

Received: 07 June 2011 / Revised: 18 October 2011 / Accepted: 07 November 2011 / Published online: 25 December 2011
© SSIEM and Springer-Verlag Berlin Heidelberg 2011

Abstract Congenital Disorders of Glycosylation (CDG)
are a group of recently described inborn errors of metabolism affecting glycosylation. CDG are disorders that have
been reported with a great variability in the clinical
presentation, especially for the most common PMM2-
CDG. The classical form is neurologic but severe forms
with multisystem disorders and *hydrops fetalis* have been
described. Here, we extend on the opposite end the clinical
spectrum to an asymptomatic PMM2-CDG case. The case
was the father of a child who died of neonatal galactosemia
few days after birth. He presented without any clinical or
biological signs, except a typical CDG 1 pattern in Western
blot of glycoproteins associated with a deficient phosphomannomutase activity in blood leukocytes and compound
heterozygosity in *PMM2* gene. The sister of the father, who
was also affected by PMM deficiency, presented with
infertility and premature ovarian failure. Finally, the

Communicated by: Eva Morava.

Competing interests: None declared.

S. Vuillaumier-Barrot (✉) · T. Dupré · C. Le Bizec · N. Seta
AP-HP, Hôpital Bichat-Claude Bernard, Laboratoire de
Biochimie, 46, rue Henri Huchard, 75877, Paris Cedex 18, France
e-mail: sandrine.vuillaumier@bch.aphp.fr

S. Vuillaumier-Barrot · T. Dupré
INSERM CRB3 U773, Faculté Xavier Bichat, 75018, Paris,
France

B. Isidor · A. David
CHU-Nantes, Service de Génétique Médicale, Nantes 44093,
France

N. Seta
Université Paris-Descartes, Faculté de Pharmacie, 75006, Paris,
France

absence of any abnormal clinical or biological signs as for
the case completes the clinical spectrum of PMM2-CDG at
its extreme end, at the opposite of the supposed total
lethality of the R141H homozygous status.

Congenital Disorders of Glycosylation (CDG) are a group
of recently described inborn errors of metabolism affecting
glycosylation. Clinically, CDG patients share many symptoms such as psychomotor retardation, ataxia, failure to
thrive, dysmorphic features, and coagulopathy, but none of
these are pathognomonic. Even more, a great variability in
the clinical presentation has been reported, especially for
PMM2-CDG (OMIM#212065), the most common up to
now CDG type (Jaeken and Carchon 2004).

PMM2-CDG is associated with deficiency in enzymatic
activity of cellular phosphomannomutase (PMM), the
enzyme responsible for conversion of Man 6-phosphate to
Man 1-phosphate, and defective *PMM2* gene. PMM2-CDG
patients present with inverted nipples, subcutaneous fat
pads, psychomotor retardation, ataxia due to cerebellar
hypoplasia, retinis pigmentosa, this phenotype still considered the classical presentation of PMM2-CDG (Grunewald
2009). However, this presentation may vary in terms of
severity and age of the patient. Here, we report on a
PMM2-CDG case without any apparent clinical symptoms.

The case was a married 38-year-old man, without any
reported health problems and normal serum transaminases,
prothrombine time, and fibrinogen values. He had a
superior education level and worked ever after graduation
in informatics. With his nonconsanguineous wife, he had
two children, first a healthy boy and then a girl. The
familial history was unremarkable until the birth of his
daughter at 37 weeks gestation after an uneventful
pregnancy. Respiratory distress appeared secondary and

the child was discharged from an intensive care unit. Then weight loss was occurred secondary to a severe secretory diarrhea and polyuria. Neurological exam showed central and peripheral hypotonia. At day 9, cardiac arrest occurred without any previous sign of cardiac failure. The only available biological sample for this child before she died was filter paper whole blood spots sent to our laboratory in order to exclude CDG, given the fact that she presented with unrelated symptoms from at least two organs and no evident diagnosis (Jaeken and Carchon 2004).

Western blotting assay of different serum glycoproteins from the daughter identified an abnormal CDG 1 glycosylation pattern. PMM activity could not be measured because of the absence of any cellular sample; therefore, we asked for parental cells. Leukocyte PMM activity for parents was not heterozygote as expected: the mother showed a normal activity and the father (case) showed a decreased one, overlapping with low heterozygote and high residual activity. As a consequence, serum glycoprotein western blot of the father was performed and also showed a typical CDG 1 glycosylation pattern.

Direct sequencing of the father's genomic DNA extracted from whole blood sample identified two mutations: c.367C>T p.R123X already described as pathogenic (Matthijs et al. 2000) and c.614A > G p.Y205C. The Y205C mutation was not found on 100 control chromosomes and after transgenesis in *E. coli* the mutated protein showed a diminished PMM activity of 25% to wild type (data not shown). His daughter was found heterozygote R123X/-.

To complete the family study (Fig 1), we obtained blood from the son, the sister, and brother of the father, all of them being apparently healthy, although the sister had been suffering from infertility and premature ovarian

failure, at 26 years old. The son and the brother were found, respectively, heterozygote, Y205C/- and R123X/-, with heterozygote PMM activities. The sister was found composite heterozygote R123X/Y205C with a deficient PMM activity and a typical CDG 1 glycosylation pattern, like the father.

In order to finalize the diagnosis for the daughter for whom the biochemical screening was positive, we tested the genes involved in secondary CDG and identified a compound heterozygosity for two frequent and severe point mutations on *GALT* (Galactose-1-P uridyltransferase; galactosemia-OMIM#230400) inherited from her parents : c.563a > G (Q188R) and c.855G>T (K285N).

In this family, as detailed in Table 1, contrarily to what was initially suspected, the daughter did not suffer from PMM2-CDG but galactosemia, and the father and his sister were the ones with PMM2-CDG.

The absence of clinical presentation in the father could be related to the mutation combination that was never described before or afterwards. However, leukocyte PMM activity was very low, just as in other PMM2-CDG patients, assessing the inborn error of metabolism. Considering the sister of the father who was also affected, the causality of PMM deficiency in infertility and premature ovarian failure she presented with, remains to be explored in a series of patients with such clinical signs as ovarian failure is a well-known feature of PMM2-CDG (Kristiansson et al. 1995).

This report is the first one about an apparently healthy subject with PMM2-CDG. The father never ever had any signs usually observed in PMM2-CDG patients, no mental delay or neurological defects, and even no other signs as reported in his medical file. His status was discovered only because of his daughter's illness; otherwise, he would have never been tested for CDG. Some patients with very mild signs have been reported previously, such as mild intellectual delay or coagulopathy. All had gone to the doctor's because of health problems and finally were screened for CDG. In a previous study, we reported *hydrops fetalis* in two related fetuses with PMM2-CDG, adding them to the list of other similar cases with CDG (Leticee et al. 2010). This unspecific sign is observed in clinically severe forms of CDG whatever the type, in presumable relation to hepatic failure commonly described in CDG.

The description of this adult PMM2-CDG case is essential. It could help understanding the discrepancy between expected (1/25,000 to 1/50,000) (Schollen et al. 2000) and observed (less than 150 PMM2-CDG patients diagnosed in France) prevalence of PMM2-CDG, which could be explained in part by fetal or neonatal death prior to diagnosis, or under ascertainment of mild and atypical cases at the other end of the spectrum. Indeed, the patients that are screened for CDG whatever their age are only the ones

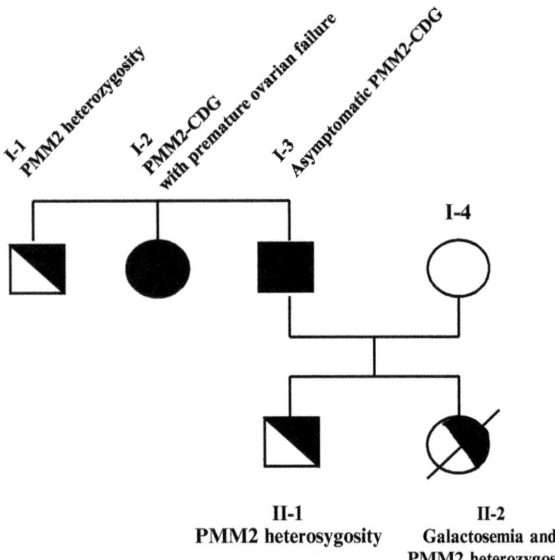

Fig. 1 Family tree for PMM2-CDG-mutated alleles

Table 1 Characteristics of the PMM2-CDG case and his family

	Leukocyte PMM activity (N > 4.2 U/g TP)	*PMM2* mutations	*GALT* mutations	CDG screening pattern	Diagnosis
I-3 Father (Case)	1.2 U/g TP	**[c.367C>T (R123X)] + [c.614A>G (Y205C)]**	[c.563a>G (Q188R)] + [N]	CDG I	Asymptomatic PMM2-CDG (GALT heterozygosity)
I-1 Brother	3.3 U/g TP	[c.367C>T (R123X)] + [N]	Not tested	Normal	PMM2 heterozygosity
I-2 Sister	1.2 U/g TP	**[c.367C>T (R123X)] + [c.614A>G (Y205C)]**	[c.563a>G (Q188R)] + [N]	CDG I	PMM2-CDG with premature ovarian failure (GALT heterozygosity)
I-4 Wife	5.1 U/g TP	[N] + [N]	[c.855G>T (K285N)] + [N]	Normal	GALT heterozygosity
II-1 Son	3.2 U/g TP	[c.614A>G (Y205C)] + [N]	Not tested	Normal	PMM2 heterozygosity
II-2 Daughter	Not tested	[c.367C>T (R123X)] + [N]	**[c.563a>G (Q188R)] + [c.855G>T (K285N)]**	CDG I	Galactosemia (PMM2 heterozygosity)

visiting specialized pediatricians, geneticists, neurologists, etc. The severity of the fetal presentation and its absence of specificity can mislead to other diagnoses than CDG, as can the presence of very mild and isolated unspecific signs, in the same way. Finally, the absence of any abnormal clinical or biological signs completes the clinical spectrum of PMM2-CDG at its extreme end, at the opposite of the supposed total lethality of the R141H homozygous status.

Synopsis

Misdiagnosis of neonatal galactosemia expands the phenotype of PMM2-CDG to an asymptomatic adult man and a woman with premature ovarian failure.

References

Grunewald S (2009) The clinical spectrum of phosphomannomutase 2 deficiency (CDG-Ia). Biochim Biophys Acta 1792:827–834

Jaeken J, Carchon H (2004) Congenital disorders of glycosylation: a booming chapter of pediatrics. Curr Opin Pediatr 16:434–439

Kristiansson B, Stibler H, Wide L (1995) Gonadal function and glycoprotein hormones in the carbohydrate-deficient glycoprotein (CDG) syndrome. Acta Paediatr 84:655–659

Leticee N, Bessieres-Grattagliano B, Dupre T et al (2010) Should PMM2-deficiency (CDG Ia) be searched in every case of unexplained hydrops fetalis? Mol Genet Metab 101:253–257

Matthijs G, Schollen E, Bjursell C et al (2000) Mutations in PMM2 that cause congenital disorders of glycosylation, type Ia (CDG-Ia). Hum Mutat 16:386–394

Schollen E, Kjaergaard S, Legius E et al (2000) Lack of Hardy-Weinberg equilibrium for the most prevalent PMM2 mutation in CDG-Ia (congenital disorders of glycosylation type Ia). Eur J Hum Genet 8:367–371

JIMD Reports
DOI 10.1007/8904_2011_115

CASE REPORT

The Ketogenic Diet Is Well Tolerated and Can Be Effective in Patients with Argininosuccinate Lyase Deficiency and Refractory Epilepsy

Rosanne Peuscher • Monique E. Dijsselhof •
Nico G. Abeling • Margreet Van Rijn •
Francjan J. Van Spronsen • Annet M. Bosch

Received: 01 August 2011 / Revised: 02 November 2011 / Accepted: 08 November 2011 / Published online: 25 December 2011
© SSIEM and Springer-Verlag Berlin Heidelberg 2011

Abstract Argininosuccinate lyase (ASL) deficiency (MIM 608310, McKusick 207900) is a rare disorder of the urea cycle, which leads to a deficiency of arginine and hyperammonemia. Epilepsy is a frequent complication of this disorder. A ketogenic diet (KD) can be a very effective therapy for refractory epilepsy, and it has been widely used in children. Until now, no experiences with the KD in patients with urea cycle defects have been reported.

We present two cases of patients with ASL deficiency and refractory epilepsy who were treated with a KD. In both patients, the KD was initiated during a hospital admission and the fat percentage of the diet was increased to above 90% in five equal steps. In patient 1, during the KD the protein intake was continued as before, and in patient 2 the natural protein was increased with 0,2 g/kg/day while the protein from the amino acid supplement (UCD-2®, Milupa) was decreased with 0,3 g/kg/day. During and after the introduction of the KD, all biochemical parameters reflecting urea cycle function and ammonia levels were stable in both patients and no signs of derangement were detected. On the KD, patient 1 demonstrated a reduction in seizure frequency of >50%, and an increase in well-being. In patient 2, no effects of the KD on the seizure frequency were noted and after 6 months the KD was discontinued.

Concluding, the KD does not cause metabolic derangement, is well tolerated, and can be effective in patients with ASL deficiency who are treated with a protein restriction.

Abbreviations

ASA	Argininosuccinic acid
ASL deficiency	Argininosuccinate – lyase deficiency
EEG	Electroencephalogram
KD	Ketogenic diet
UCD	Urea cycle disorder

Communicated by: Claude Bachmann.

Competing interests: None declared.

R. Peuscher · A.M. Bosch (✉)
Department of Pediatrics, Academic Medical Center (H7-270),
University of Amsterdam, Meibergdreef 9,
1105 AZ, Amsterdam, The Netherlands
e-mail: a.m.bosch@amc.nl

M.E. Dijsselhof
Department of Dietetics, Academic Medical Center, University
of Amsterdam, Amsterdam, The Netherlands

N.G. Abeling
Laboratory Genetic Metabolic Diseases, Academic Medical
Center, University of Amsterdam, Amsterdam, The Netherlands

M. Van Rijn · F.J. Van Spronsen
Department of Pediatrics, University Medical Center Groningen,
Groningen, The Netherlands

Introduction

We present two patients with a urea cycle defect and refractory epilepsy. The usual treatment consists of supplementation of L-arginine, the missing enzyme product; restriction of natural protein intake to reduce the nitrogen load, often with a part replaced by special essential amino acid mixtures in order to avoid their limitation of protein synthesis; if needed drugs to stimulate alternate pathways of nitrogen excretion. Treatment of epilepsy of these patients is hampered by the relative contraindication of valproate as this may increase the ammonia level and may give rise to liver damage. Due to the fact that the epilepsy was difficult to treat, a ketogenic diet (KD) was initiated in both patients.

Patient 1

A now 5-year-old boy presented with hyperammonemia in the first 48 h of life. The maximum ammonia level was 1,085 μmol/l. Treatment consisting of peritoneal dialysis and administration of sodium benzoate and arginine was started immediately, and ammonia levels normalized within 24 h. Urine and plasma argininosuccinic acid (ASA) and plasma citrulline and glutamine were strongly elevated, and mutation analysis demonstrated heterozygosity for a c.857 A.G (Q286R) and a c.447-1 G > A mutation in the argininosuccinate lyase gene (ASL,EC 4.3.2.1), confirming the diagnosis argininosuccinate lyase deficiency (ASL deficiency, MIM 608310, McKusick 207900).

After the serious clinical presentation in the first week of life and during treatment with a protein restricted diet, sodium phenylbutyrate and arginine, no further episodes of metabolic decompensation occurred and the patient demonstrated normal psychomotor development until he developed severe and refractory epilepsy at the age of 18 months. In the following years, his epilepsy did not respond to levetiracetam, carbamazepine, clobazam, or ethosuximide. Severe seizures necessitated the use of diazepam on average twice weekly. His development was fully arrested after the start of the epilepsy. The electroencephalogram (EEG) showed almost continuous presence of high-voltage, specific epileptiform discharges. Cerebral MRI at age 2 years and 8 months demonstrated bilateral loss of tissue at the nucleus caudatus and atrophy of the frontal cortex.

Patient 2

This male patient presented on the third day of life with hyperammonemia resulting in seizures and coma. The highest ammonia level was 847 μmol/l, which rapidly normalized after treatment with sodium benzoate, sodium phenylacetate, arginine, and carnitine. The diagnosis of argininosuccinate lyase deficiency was made based on the amino acid profile, with strongly elevated glutamine and very low arginine levels and a clearly increased argininosuccinic acid in the urine. In this patient, milestones were normal at 12 months of age notwithstanding some hypertonia. At the age of 2.5 years, he developed severe refractory epilepsy, and during the following years he demonstrated little development. He developed a status epilepticus requiring high-dose midazolam and artificial ventilation at the age of 7 years. Cerebral MRI at ages 6 months and 2 years demonstrated no abnormalities, and an MRI at age 7 years and 8 month, just before the start of the ketogenic diet, demonstrated loss of tissue at the nucleus caudatus and minor myelinisation abnormalities.

Introduction of the Ketogenic Diet

Patient 1 was four years old, his weight was 17.5 kg, and he was fed by gastrostomy at the start of the KD. He had a daily intake of 1.0 g per kg of natural protein and 0.4 g per kg of protein from an essential amino acid supplement (UCD-2® Milupa) to meet his total requirements of essential amino acids. Energy intake was adequate for his age (1,350 calories per day, divided into five portions of tube-feeding) and the dietary fat intake was normal with 40% of the total energy amount. He was treated with sodium benzoate 1,200 mg four times per day (274 mg/kg/day) and L-Arginine 1,500 mg four times per day (324 mg/kg/day). The patient was admitted to the hospital for the introduction of the KD. In the KD of this patient, the protein intake was continued as before. With the use of Ketocal® (Nutricia) and a fat emulsion (Calogen® Nutricia), the amount of fat in the diet was raised to a total of 90% in five equal steps.

During the introduction of the KD, the total amount of calories per day was raised with 200 calories per day because of weight loss during the introduction of the KD (and to avoid catabolism c.q. protein breakdown as a consequence). With a total of 90% fat in the diet, the patient reached the state of ketosis. This enteral feeding was well tolerated except for diarrhea in the first days for which a fiber supplement was added to the feeding.

Patient 2 was 7 years old with a weight of 25,6 kg at the start of the KD. He had a daily intake of 0.8 g per kg of natural protein and 0.5 g per kg of protein from an essential amino acid supplement (UCD-2®, Milupa) (UCD2®) to meet his total requirements of essential amino acids. Energy intake was adequate for his age (1,875 kcal per day distributed over six meals in a combination of tube feeding per gastrostomy and normal meals), and the dietary fat intake was normal with 34% of the total energy amount. He was treated with Arginine HCl 3,900 mg four times per day (600 mg/kg/day).

In the KD of this patient, the protein intake was 1 g of natural protein and 0.2 g per kg from an essential amino acid supplement (UCD-2®, Milupa). With the use of Ketocal®, the amount of fat in the diet was raised to a total of 92% in five equal steps. The diet was started as total tube feeding in a period of epileptic decompensation, then after clinical improvement, partly changed to ketogenic normal oral meals equivalent in fat, protein, and energy.

With a total of 92% fat in the diet, the patient reached only moderate ketosis (1-2+ in urine). LCT fat was partly replaced by MCT fat (3 ml Liquigen® equal to 1,5 MCT fat per kg b.w.), which resulted in a somewhat higher ketosis rate (1-3+ in urine). This feeding was well tolerated and after 3 months reduced in energy with 10% because of weight gain above the usual growth rate. Medication was thoroughly checked for carbohydrate content.

Safety and Effect

During and after the introduction of the KD in both patients all biochemical parameters reflecting urea cycle function and ammonia levels were stable, and no signs of derangement were detected. Levels of essential amino acids were within the normal range, with a slight decrease of alanine and an increase of branched chain amino acids, which may result from the increased gluconeogenesis during the KD (Table 1.) All measurements were in fed state due to the frequency of the feedings. Blood glucose concentrations remained within aimed range (> 4 mmol/L). Measurements of bicarbonate in patient 1 and CO_2 in patient 2 demonstrated no metabolic acidosis during ketosis.

Before the initiation of the KD, patient 1 demonstrated an average of 0.5–3 seizures per day and diazepam was administered on average twice weekly. A decrease of the frequency of seizures was noted after the first week of introduction of the KD. In the 3 months period after introduction of the KD the patient demonstrated a 50% reduction in seizure frequency. The administration of diazepam for severe seizures is still necessary twice weekly. However, with the decrease in seizure frequency there is a remarkable improvement in his well-being. He is more alert and now communicates more actively with his parents.

In patient 2, no effects on seizure frequency were noted during the application of the KD and after 6 months the KD was discontinued and the enteral feeding was gradually converted into his original feeding in five equal steps.

Discussion

ASL deficiency is a rare autosomal recessive disorder of the urea cycle caused by the deficiency of the enzyme argininosuccinate lyase, which leads to a deficiency of arginine (and fumarate/malate in the cytosol) and to hyperammonemia. Patients with ASL deficiency may present with either a severe neonatal form or a late onset form (Erez et al. 2011. Epilepsy is a frequent complication of ASL deficiency, both in patients with a clinical presentation and in those detected with newborn screening (Ficicioglu et al. 2009; Grioni et al. 2011).

A KD can be a very effective therapy for refractory epilepsy (Bough 2008; Kossoff et al. 2009), and it has been widely used in children. It was first formulated by Wilder (1921), however only frequently used since the 1990s. It has been known since the time of Hippocrates that fasting is an effective treatment for seizures, and the KD was designed to mimic the fasting state. Despite intensive research in recent years, the mechanism by which the diet protects against seizures remains unknown. Recent hypotheses state that while epileptic foci are hypometabolic areas, metabolic derangement leads to synaptic instability and the development of seizures. Initiation of KD leads to upregulation of several pathways, including energy metabolism genes, mitochondrial biogenesis and an increase in energy reserves which results in more resistant brain tissue to metabolic stress and an increase in seizure threshold (Bough 2008).

To the best of our knowledge, no reports of the use of a KD in urea cycle defects could be found. Furthermore, a question posted on the Metab-L, a mailing list on inborn errors of metabolism (http://www.daneel.franken.de/metab-l) did result in negative advice but no experiences were reported. Until now, the effects of the KD have only been described in other metabolic disorders leading to refractory epilepsy (Klepper 2008). Because of the very poor quality of life due to the refractory epilepsy, a trial with a KD was started in both patients. Patients with urea cycle defects are prone for decompensation when they are fasting, due to protein catabolism exceeding its synthesis. While KD induces a

Table 1 Plasma amino acids before start KD and after reaching ketosis

		Glutamine ref 373–709 µmol/l	Arginine ref 38–98 µmol/l	Alanine ref 158–314 µmol/l	Valine ref 133–73 µmol/l	Leucine ref 64–164 µmol/l	Isoleucine ref 3 1–83 µmol/l
Pat 1 6 months before start KD	n mean (range)	$N=7$ 541 (466–635)	$N=7$ 79 (30–119)	$N=7$ 344 (229–485)	$N=7$ 141 (112–181)	$N=6$ 64 (48–88)	$N=7$ 33 (15–52)
Pat 1 6 months after reaching ketosis	n mean (range)	$N=8$ 565 (432–718)	$N=8$ 58 (22–201)	$N=8$ 152 (75–225)	$N=8$ 227 (126–288)	$N=8$ 113 (41–173)	$N=8$ 74 (27–113)
Pat 2 6 months before start KD	n mean (range)	$N=3$ 516 502–538	$N=2$ 80 72–87	$N=2$ 341 306–376	$N=2$ 227 203–250	na	$N=2$ 65 (48–82)
Pat 2 6 months after reaching ketosis	n mean (range)	$N=3$ 460 326–517	$N=3$ 60 29–106	$N=3$ 132 103–156	$N=3$ 224 186–264	$N=1$ 109	$N=3$ 70 61–85

state of ketosis, it is due to administration of a high fat diet and not to catabolism. No signs of metabolic decompensation were detected in our patients.

In general, numerous adverse effects are seen in KD including weight loss, gastroesophageal reflux, constipation, and diarrhea (Coppola et al. 2009). Diarrhea is seen in approximately 13% of patients receiving KD (Neal et al. 2008). Side effects in patient 1 were indeed diarrhea, and gastroesophageal reflux for which he was treated with omeprazole. Also, he suffered from calciuria necessitating oral potassium citrate. However, the positive effects of the KD on his well-being outweigh the adverse affects by far.

Conclusion

We conclude that the KD does not cause metabolic decompensation, is well tolerated and can be effective in patients who suffer from ASL deficiency and who are treated with protein restriction. The KD may well be an acceptable option for patients with refractory epilepsy due to other inborn errors of metabolism.

References

Bough K (2008) Energy metabolism as part of the anticonvulsant mechanism of the ketogenic diet. Epilepsia 49(Suppl 8):s91–s93

Coppola G, Verrotti A, Ammendola E, Operto FF, Corte RD, Signoriello G, Pascotto A (2009) Ketogenic diet for the treatment of catastrophic epileptic encephalopathies in childhood. Eur J Paediatr Neurol 14(3):229–234

Erez A, Nagamani SCS, Lee B (2011) Argininosuccinate lyase deficiency Argininosuccinic aciduria and beyond. Am J Med Genet C Semin Med Genet 157:45–53

Ficicioglu C, Mandell R, Shih VE (2009) Argininosuccinate lyase deficiency: longterm outcome of 13 patients detected by newborn screening. Mol Genet Metab 98(3):273–277

Grioni D, Furlan F, Corbetta C, Barboni C, Lastrico A, Marzocchi GM, Contri M, Gamba A, Vizziello P, Parini R (2011) Epilepsy and argininosuccinic aciduria. Neuropediatrics 42(3):97–103

Klepper J (2008) Glucose transporter deficiency syndrome (GLUT1DS) and the ketogenic diet. Epilepsia 49(Suppl 8): s46–s49

Kossoff EH, Zupec-Kania BA, Amark PE et al (2009) Optimal clinical management of children receiving the ketogenic diet: recommendations of the international ketogenic diet study group. Epilepsia 50(2):304–317

Neal EG, Chaffe H, Schwartz RH et al (2008) The ketogenic diet for the treatment of childhood epilepsy: a randomised controlled trial. Lancet Neurol 7(6):500–506

Wilder RM (1921) The effect of ketonemia on the course of epilepsy. Mayo Clin Bulletin 2:307–308